INTERNATIONAL SERIES ON
MATERIALS SCIENCE AND TECHNOLOGY
GENERAL EDITOR: D. W. HOPKINS

VOLUME 13

Engineering Calculations in Radiative Heat Transfer

Engineering Calculations in Radiative Heat Transfer

W. A. GRAY AND R. MÜLLER

PERGAMON PRESS

OXFORD · NEW YORK · TORONTO
SYDNEY · BRAUNSCHWEIG

Pergamon Press Ltd., Headington Hill Hall, Oxford
Pergamon Press Inc., Maxwell House, Fairview Park, Elmsford,
New York 10523
Pergamon of Canada Ltd., 207 Queen's Quay West, Toronto 1
Pergamon Press (Aust.) Pty. Ltd., 19a Boundary Street,
Rushcutters Bay, N.S.W. 2011, Australia
Vieweg & Sohn GmbH, Burgplatz 1, Braunschweig

First edition 1974

Library of Congress Cataloging in Publication Data
Gray, William Alan
Engineering calculations in radiative heat transfer
(International series on materials science and technology, v. 13)
Bibliography: p.   Heat—Radiation and absorption. I. Müller
Rudolph, joint author. II. Title.
QC338.G7 1974   536'.33   73–17321
ISBN 0–08–017786–7
ISBN 0–08–017787–5 (flexicover)

Printed in Great Britain by A. Wheaton & Co., Exeter

Contents

Nomenclature

A	area
A_m	projected area per unit mass of particle material
A_n	projected area per particle
A_v	projected area per unit volume of particle material
C_{CO_2}	correction factor for the emissivity of carbon dioxide
C_{H_2O}	correction factor for the emissivity of water vapour
D	diameter;
	characterizing dimension
D_z	scattered solar radiation
D_{12}	fraction of energy radiated from a black body within the wavelength range λ_1 to λ_2
D^*	normalized detectivity
E	radiative flux emitted by a body
E_b	radiative flux emitted by a black body (black-body emissive power)
E_λ	monochromatic radiative flux emitted by a body
$E_{b\lambda}$	monochromatic black-body emissive power
F_{ab}	relative absorption cross-sectional area
F_{at}	relative attenuation cross-sectional area
F_{sc}	relative scattering cross-sectional area
F_{12}	view factor from surface 1 to surface 2
G	solar radiation
G_g	global solar radiation (direct + diffuse)
G_m	solar radiation attenuated by passage through a turbid atmosphere of relative air mass m
G_n	solar radiation incident upon a surface normal to that radiation
G_0	solar constant
G_t	solar radiation arriving at the Earth (neglecting attenuation)
G_λ	monochromatic solar radiation

GHA	Greenwich hour angle
H	radiative flux incident upon a surface;
	hour angle between noon and sunrise;
	enthalpy
I	intensity of radiation
I_n	intensity of radiation normal to a surface
K	attenuation or extinction coefficient
L	length
L_m	mean beam length
L_{12}	mean beam length between surface (or medium) 1 and surface (or medium) 2
M	molecular weight
NEP	noise equivalent power
P	total pressure
Q	rate of heat transfer
R	random number;
	gas constant;
	electrical resistance
R_a	aphelion distance
R_p	perihelion distance
S_v	surface area of particles per unit volume of particle material
T	absolute temperature;
	turbidity factor
T_b	effective black body (or brightness) temperature
V	volume
W	radiative flux leaving a surface (radiosity)
$W(T)$	Wien approximation for E_b
$\overline{12}$	exchange area between two surfaces (1 and 2)
a	altitude (90°—zenith angle);
	weighting coefficient for grey gas fit
az	azithumal angle
c	specific heat;
	velocity of light
c_m	mass concentration of particles in suspension
c_n	number of particles per unit volume of suspension
c_v	volume concentration of particles in suspension

c_1, c_2 constants in Planck's expression

d declination;

particle diameter

e eccentricity of the elliptical path of the Earth

f ratio of the mean beam length to the characterizing dimension (L_m/D)

f_m average value of L_m/D

$\overline{g_1 g_2}$ exchange area between two gases (1 and 2)

$\overline{g_1 s_2}$ exchange area between a gas (1) and a surface (2)

h convective heat transfer coefficient;

local hour angle

k attenuation coefficient used in conjunction with partial pressure $(k = K/p)$

k' absorption index

l latitude;

length

m relative air mass;

complex optical property

m_a mass of air

m_f mass of fuel

n refractive index;

angle

p partial pressure

q rate of heat transfer per unit area

r distance;

radius

$\overline{s_1 s_2}$ exchange area between two surfaces (1 and 2)

t time

t_a transmission coefficient through the atmosphere for unit mass of air

w solid angle

z zenith angle

α absorptivity

α_λ monochromatic absorptivity

ϵ emissivity

ϵ_g emissivity of a gas

ϵ_n	emissivity in a direction normal to the surface
ϵ_λ	monochromatic emissivity
η	efficiency;
θ	circumferential angle
λ	wavelength
λ_{max}	wavelength at which maximum flux is emitted
ν	frequency
ρ	density; reflectivity
ρ_f	density of fuel
ρ_p	density of products of combustion
ρ_λ	monochromatic reflectivity
σ	Stefan-Boltzmann constant
τ	transmissivity
τ_λ	monochromatic transmissivity
ϕ	angle between direction of emission and normal to the surface (cone angle)
Ω	angular velocity of the Earth

Acknowledgements

WE ARE very grateful for the help given to us by Dr. B. A. Lilley, colleagues in the Department of Fuel and Combustion, Anne Müller and Pauline Gray.

CHAPTER 1

Basic Principles of Thermal Radiation

1.1. NATURE OF RADIATION

Radiation is the transmission of energy by electromagnetic waves; the energy transmitted is called radiant energy. However, the term radiation is also commonly used to describe the radiant energy itself. Electromagnetic waves are characterized by their wavelength or frequency, frequency being inversely proportional to wavelength. Wavelength is usually used in radiative heat transfer analysis.

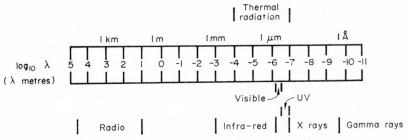

FIG. 1.1. Spectrum of electromagnetic waves.

Figure 1.1 shows the electromagnetic spectrum and the names given to radiation transmitted in various ranges of wavelengths. The nature of radiation and its transport are not fully understood but they can be described satisfactorily either by wave or quantum theory. In simple terms, radiation travels in space with the velocity of light and does not require the presence of an intervening medium for its propagation. The velocity of light (c) is the constant of proportionality which relates wavelength (λ) and frequency (ν):

$$\lambda = c/\nu. \tag{1.1}$$

1

The frequency of radiation depends on the nature of the source. For example, a metal bombarded by high-energy electrons emits X-rays, high-frequency electric currents generate radio waves and a body emits thermal radiation by virtue of its temperature.

Radiation in the wavelength range 0·1 to 100 μm (micrometre or micron), when incident upon a body, will heat it and, consequently, is called thermal radiation. In addition, since radiation within the wavelength band of 0·38 to 0·76 μm affects the optic nerves, we can see thermal radiation within this band as light. In practice, this means that we cannot see the thermal radiation from a body below about 500°C but we can feel the heat radiated from it.

Thermal radiation, like the other forms of electromagnetic radiation, can often be considered to travel in straight lines in a uniform medium. (Departures from this simple assumption will be dealt with later.) Consequently, opaque bodies cast shadows when placed in the path of thermal radiation and one body cannot receive radiation directly from another unless it can "see" it.

All bodies at a temperature above absolute zero emit thermal radiation and, consequently, lose energy. If it were possible to isolate a body completely, it would continue to radiate and lose energy until its own temperature had fallen to absolute zero. Complete isolation is not possible and the body is heated by its surroundings (usually by all three mechanisms of heat transfer, namely, conduction, convection and radiation).

If a body is placed in surroundings at the same temperature as itself, its temperature does not change. Nevertheless, it continues to radiate energy and, simultaneously, receive energy at the same rate from its surroundings. This concept is called Prevost's principle of exchange.

1.2. ABSORPTION, REFLECTION AND TRANSMISSION

When radiation falls upon a body (solid, liquid or gas), a fraction (α) of it is absorbed, a fraction (ρ) is reflected and the remainder (τ) is transmitted through the body:

$$\alpha + \rho + \tau = 1. \qquad (1.2)$$

α is called the absorptivity, ρ the reflectivity and τ the transmissivity.

Most solid materials absorb practically all radiation within a very thin surface layer, less than 1 mm thick. For these opaque materials, $\tau = 0$ and

$$a + \rho = 1. \tag{1.3}$$

Certain solids and liquids transmit radiation at specific wavelengths unless they are very thick. These materials (glass, inorganic crystals, etc.) are transparent to radiation only at these wavelengths; at other wavelengths they are opaque to radiation. Thus, ordinary clear glass is transparent in the visible range and in the infra-red up to 2·5 μm, whereas very thin slices of semiconductor materials such as silicon and germanium are opaque in the visible region but transparent over parts of the infra-red region beyond 1·0 and 1·8 μm respectively.

The surface of a solid which is highly polished and smooth behaves like a mirror to thermal radiation, that is, the angle of reflection equals the angle of incidence. In this case, reflection is called regular or specular. Most industrially important materials possess rough surfaces, that is, their surface irregularities are large compared to the wavelength of radiation. Reflection of radiation from this kind of surface occurs indiscriminately in all directions and is called diffuse.

1.3. BLACK-BODY RADIATION

A visibly matt black surface absorbs all the light which falls upon it. By analogy, a surface which absorbs all the thermal radiation incident upon it is also called a black surface and a body which absorbs all incident radiation is a black body. In practice, the perfectly black body or surface does not exist but many approach it or can be made to approach it. For example, the surface of a body may be coated with carbon black to produce a near-black surface as far as thermal radiation is concerned. In any case, the concept of a black body or surface is very useful in the analysis of radiative heat transfer.†

Consequently, we can relate the behaviour of material under examination to that of a black body in similar circumstances or we can ensure that the material is placed in a situation in which it behaves as a black body.

---

† Since most solids absorb thermal radiation very close to the surface, it is usual to refer to the surface of these bodies and not to the bodies themselves.

A black body emits the maximum possible amount of radiation corresponding to its temperature; other bodies emit less. Summarizing, a black body can be defined either as a body which

(a) absorbs all radiation incident upon it (and reflects and transmits none) or
(b) emits, at any particular temperature, the maximum possible amount of thermal radiation.

1.4. STEFAN–BOLTZMANN LAW

The rate at which energy is radiated from a black body is proportional to the fourth power of its absolute temperature

$$E_b = \sigma T^4, \tag{1.4}$$

where E_b is the rate at which energy is radiated from unit area of the surface of a black body at an absolute temperature T to the hemisphere of space above it (black-body emissive power) and σ is the Stefan–Boltzmann constant ($56\cdot7 \times 10^{-12}$ kW/m^2 K^4). For calculations, it is convenient to use eq. (1.4) in the form

$$E_b = 56\cdot7 \times (T/1000)^4 \text{ kW/m}^2 \text{ K}^4. \tag{1.5}$$

EXAMPLE 1.1

Calculate the rate at which energy is radiated from a black surface, $0\cdot3$ m^2 in area, which is (a) red hot at a temperature of 700°C (973 K) and (b) white hot at 1500°C (1773 K).

From the Stefan–Boltzmann law and $Q = E_b A$:

(a) $Q = \sigma T^4 A$
$= 56\cdot7 \times (973/1000)^4 \times 0\cdot3$
$= 15\cdot2$ kW,

(b) $Q = 168$ kW.

1.5. VARIATION OF THERMAL RADIATION WITH WAVE-LENGTH—PLANCK'S DISTRIBUTION LAW

Figure 1.2 shows the radiative flux (energy per unit area per unit time) emitted by a black body as a function of wavelength for a number

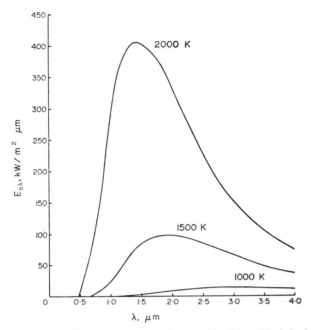

FIG. 1.2. Spectral distribution of radiation emitted by a black body.

of temperatures. These curves give the maximum possible radiation that can be emitted by any body for the temperatures shown. The function, which Planck derived from the quantum theory, is

$$E_{b\lambda} = \frac{c_1 \lambda^{-5}}{\exp(c_2/\lambda T) - 1} \qquad (1.6)$$

where $E_{b\lambda}$ is the monochromatic emissive power of a black body, λ the wavelength, T the absolute temperature and c_1 and c_2 are constants equal to $3 \cdot 74 \times 10^{-19}$ kW m^2 and $1 \cdot 439 \times 10^{-2}$ m K respectively.

The radiation emitted over the whole range of wavelengths may be found by integrating Planck's expression from 0 to ∞, to give σT^4.

Since a unique curve is produced for each temperature (Fig. 1.2), Planck's law may be presented more conveniently as a single curve by plotting $E_{b\lambda}/T^5$ against λT (Fig. 1.3).

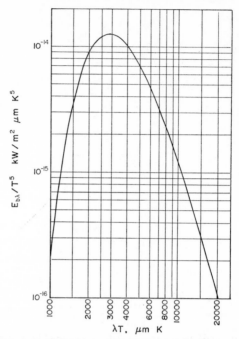

FIG. 1.3. Distribution of radiation emitted by a black body.

1.6. WIEN'S DISPLACEMENT LAW

The curve of radiative flux versus wavelength has the same form for every temperature but as the temperature is increased the height of the curve increases and the maximum moves towards the shorter wavelengths (Fig. 1.2). If the temperature of a hot body is less than 500°C (773 K), virtually none of the radiation will fall within the band of wavelengths corresponding to visible light. If the temperature of the body is increased, some radiation will fall within the visible range and, at about 700°C (973 K), the surface glows dull red. With further increase in temperature, the colour changes to cherry red at 900°C (1173 K), orange red at 1100°C (1373 K) and, finally, at temperatures greater than about 1400°C (1673 K), when sufficient energy is emitted in the visible range, the body becomes white hot. At the same time, the total quantity of heat radiated, which is proportional to T^4, increases rapidly. Even at 2500°C (2773 K), the temperature of the element of the incandescent lamp, only about 10 per cent of the energy is emitted in the visible range, which illustrates that the incandescent lamp is a more efficient source of heat than light (see Example 1.3).

The wavelength at which maximum flux is emitted (λ_{max}) is inversely proportional to the absolute temperature, a relationship known as Wien's displacement law:

$$\lambda_{max}T = 2.9 \times 10^{-3} \text{ m K} = 2900 \text{ } \mu\text{m K.} \tag{1.7}$$

EXAMPLE 1.2

Determine the wavelengths at which maximum radiant energy is emitted from surfaces at temperatures in the range 500 to 6000 K and plot these values.

The solution to this example, obtained from eq. (1.7), is presented graphically in Fig. 1.4.

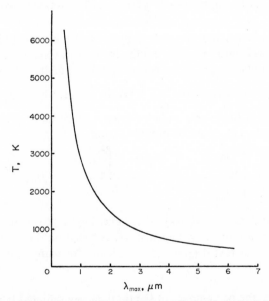

FIG. 1.4. Wavelengths at which maximum radiation is emitted by a black body.

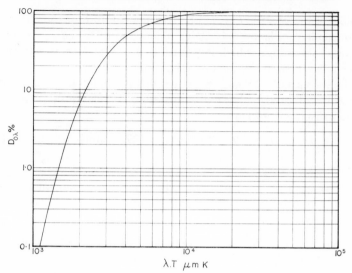

FIG. 1.5. Fraction of black-body radiation emitted between wavelengths 0 and λ.

Many problems involve an estimate of the energy radiated at a specified wavelength or within a finite band of wavelengths. The energy emitted by a black body in the wavelength band defined by λ_1 and λ_2 is $\int_{\lambda_1}^{\lambda_2} E_{b\lambda} d\lambda$, or the area under the curve in Fig. 1.2 between λ_1 and λ_2. The fraction of the energy radiated from a black body over this range of wavelengths (D_{12}) can be obtained from

$$D_{12} = \frac{\int_{\lambda_1}^{\lambda_2} E_{b\lambda} d\lambda}{\int_0^{\infty} E_{b\lambda} d\lambda} = \frac{\int_0^{\lambda_2} E_{b\lambda} d\lambda - \int_0^{\lambda_1} E_{b\lambda} d\lambda}{\sigma T_b^4} = D_{02} - D_{01}. \quad (1.8)$$

Tabulated data (Pivovonsky, 1961) may be used to evaluate eq. (1.8), and Fig. 1.5, a plot of $D_{0\lambda}$ against λT, is based on data selected from Table III of that text. Figure 1.5 is sufficient for many applications.

EXAMPLE 1.3

Determine the percentages of the total energy radiated from surfaces at 1000, 2000, 3000 and 6000 K which lie in the visible and infra-red regions of the spectrum. These temperatures correspond approximately to those of the bar of a domestic electric fire, the theoretical maximum flame temperatures of methane in air and in oxygen and the temperature of the Sun respectively.

The values of λT for these wavelengths and temperatures and the corresponding quantities of radiation determined from Fig. 1.5 are shown in Table 1.1.

TABLE 1.1

Temperature, K	$\lambda = 0\cdot38\ \mu m$		$\lambda = 0\cdot76\ \mu m$		$\lambda = 1000\ \mu m$	
	λT $\mu m\ K$	$D_{0\lambda}$ %	λT $\mu m\ K$	$D_{0\lambda}$ %	λT $\mu m\ K$	$D_{0\lambda}$ %
1000	380	<0·1	760	<0·1	10^6	100
2000	760	<0·1	1520	1·5	2×10^6	100
3000	1140	0·14	2280	11·5	3×10^6	100
6000	2280	11·5	4560	57·0	6×10^6	100

Consequently, the percentages of energy present in the visible and infrared parts of the spectrum are those in Table 1.2

TABLE 1.2

Temperature, K	Energy, per cent	
	Visible (0·38–0·76 μm)	Infra-red (0·76–1000 μm)
1000	<0·1	>9·99
2000	1·5	98·5
3000	11·4	89·5
6000	45·5	43·0

1.7. INTENSITY OF RADIATION

So far, we have been concerned with the total amount of radiation emitted from a surface into all the space above it (kW/m^2). It is also necessary to consider the intensity of radiation (I) emitted through specific solid angles (kW/m² ster). It can be shown that the intensity of radiation normal to the surface is related simply to the black-body emissive power by

$$I_n = E_b/\pi. \tag{1.9}$$

1.8. LAMBERT'S LAW

Equation (1.9) defines the intensity of radiation from a surface of unit area, normal to the direction of emission. When the direction of emission is at an angle ϕ to the normal to the surface (Fig. 1.6a), the projected area of emission is $\cos \phi$ and not 1. In this case, the intensity of emission is

$$I = E_b \cos \phi/\pi. \tag{1.10}$$

Equation (1.10) is a statement of the Lambert cosine law and surfaces which behave in this way are often known as Lambert surfaces. A black surface is also a Lambert surface.

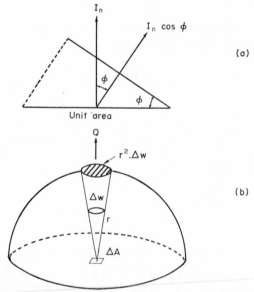

FIG. 1.6. Intensity of radiation emitted by a surface.

1.9. INVERSE SQUARE LAW

Consider the small cone of radiation emitted from a surface of a small area (ΔA) and transmitted through the solid angle Δw (Fig. 1.6b). It intercepts an area $r^2\Delta w$ on a sphere of radius r, which has its centre at the source of radiation. Since this area is proportional to r^2 then the radiation received per unit area is inversely proportional to r^2, a relationship known as the inverse square law.

The radiation from a small area of surface (ΔA) emitted through a small solid angle (Δw) in a direction normal to the surface is

$$Q = I\Delta A\Delta w. \tag{1.11}$$

And that emitted in a direction ϕ to the normal is

$$Q = I\Delta A \cos \phi\Delta w. \tag{1.12}$$

These equations do not apply to non-diffuse (specular) emitters.

EXAMPLE 1.4

Radiation from the Sun is incident upon a small surface on the Earth, the normal to the surface making an angle of 30° with the direction of solar radiation. If the radiation is totally and diffusely reflected, calculate the amount of energy reflected from 1 square metre of surface. The intensity of solar radiation is $64100/\pi$ kW/m² ster.

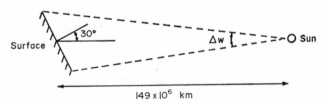

FIG. 1.7. Reflected solar radiation.

Figure 1.7 illustrates the situation described in Example 1.4. Since the angle subtended by the Earth at the Sun is small, the quantity of radiation, Q_i, incident on unit surface normal to that radiation, is

$$Q_i = A_sI_s\Delta w$$

where I_s is the effective intensity of solar radiation from a disc of area A_s and $\Delta w = 1\cdot0/(149 \times 10^6)^2$.

For a surface whose normal is inclined at an angle of 30° to the radiation

$$Q_i = \frac{\pi(1 \cdot 39 \times 10^6)^2 \times 0 \cdot 866 \times 64100 \times 1 \cdot 0}{4\pi(149 \times 10^6)^2.} = 1 \cdot 26 \text{ kW.}$$

Since total reflection occurs, Q_i is also the amount of energy reflected. It corresponds to emission from a black body at a temperature of 386 K.

1.10. ABSORPTIVITY OF A REAL SURFACE

The absorptivity of a real body or surface is not the same for all wavelengths and all angles of incidence. Consequently, it is usual to define the monochromatic absorptivity at a particular wavelength (λ) and angle of incidence (ϕ) as the fraction of the energy absorbed by the surface to that absorbed by a black body or surface. It is usually found that the absorptivity defined in this way is also a function of temperature:

$$\alpha = \alpha(\lambda, \phi, T). \tag{1.13}$$

Reflectivity and transmissivity (ρ and τ) also have monochromatic counterparts.

1.11. EMISSIVITY OF A REAL SURFACE

The emissivity of a real body or surface is defined as the ratio of the radiative flux (E) emitted by the body to that (E_b) emitted by a black body at the same temperature [eq. (1.14)]. Again for solid bodies, it is usual to refer to surface emissivities

$$\epsilon = E/E_b. \tag{1.14}$$

Emissivities can be defined at specific wavelengths (monochromatic emissivities) and, for non-Lambert surfaces, should be defined for specific angles of emission:

$$\epsilon(\lambda, \phi) = E(\lambda, \phi)/E_b(\lambda). \tag{1.15}$$

Again, the monochromatic emissivity, like monochromatic absorptivity, is usually found to be a function of surface temperature as well as wavelength and angle of emission. In practice, it is often assumed that there is little variation of emissivity with angle of emission.

1.12. GREY BODIES

A body whose emissivity is constant with wavelength is known as a grey body and many materials approximate to this ideal case. A material whose monochromatic emissivity is not constant with wavelength, angle of emission and surface temperature is called a selective emitter. Figure 1.8 shows a grey body distribution of radiation for an emissivity of 0·8. The energy radiated from a grey body is

$$E = \epsilon\sigma T^4. \qquad (1.16)$$

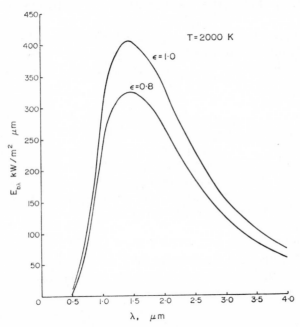

FIG. 1.8. Spectral distribution of radiation emitted by a grey body.

EXAMPLE 1.5

The temperature of a 1-kW electric fire bar is 850°C (1123 K). Calculate the radiant efficiency of the fire bar if it is 0·3 m long, 10 mm in diameter and can be considered as a grey body with an emissivity of 0·92.

The total quantity of radiative energy emitted by the bar is

$$Q = A\epsilon\sigma T^4$$
$$= 0\cdot3 \times \pi \times 0\cdot01 \times 0\cdot92 \times 56\cdot7 \times (1123/1000)^4$$
$$= 0\cdot78 \text{ kW}.$$

Since the total energy supplied to the bar is 1 kW, the radiant efficiency is 78 per cent.

1.13. EFFECTIVE BLACK-BODY (OR BRIGHTNESS) TEMPERATURE

The effective black-body temperature (T_b) of a body is determined by measuring the radiative flux leaving the body ($\epsilon\sigma T^4$ for a grey body) and equating it to that which would leave a black body at a temperature T_b:

$$\epsilon\sigma T^4 = \sigma T_b^4. \tag{1.17}$$

Consequently, as far as the total quantity of radiation emitted by a grey surface is concerned, it is not necessary to know its temperature and emissivity but merely its effective black-body (or brightness) temperature. This temperature is often determined by measuring the radiative flux from the surface with an instrument which has been calibrated by means of a black-body source.

EXAMPLE 1.6

Calculate the true temperature of a domestic electric fire bar of emissivity 0·92 if the effective black-body (or brightness) temperature has been measured as 825°C (1098 K).

$$T_b = 1098 \text{ K},$$
$$\sigma T_b^4 = \epsilon\sigma T^4.$$

Therefore $\quad T = T_b/(0\cdot92)^{1/4}$
$$= 1120 \text{ K or } 847°C.$$

Although the overall quantities of radiative flux are set equal in eq. (1.17), the spectral radiation from a grey body is not the same as

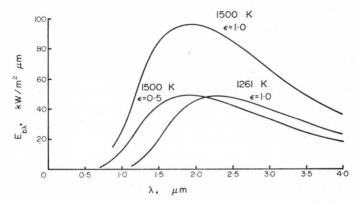

FIG. 1.9. Spectral distribution of radiation emitted by a black body at the brightness temperature of a grey body.

from a black body at its effective black-body temperature. Figure 1.9 illustrates this point by showing the spectral emission from a grey body of emissivity 0·5 and at a temperature of 1500 K and from black bodies at 1500 and 1261 K, the last being the effective black-body temperature of the grey body.

1.14. KIRCHHOFF'S LAW

Kirchhoff's law may be stated as: the monochromatic emissivity of a surface is equal to its monochromatic absorptivity irrespective of the difference in temperatures corresponding to the emitted (T_1) and incident (T_2) radiation.

$$\epsilon_\lambda = a_\lambda. \qquad (1.18)$$

Since this law is valid at all wavelengths (λ) and angles (ϕ), it may be shown (Appendix 1) that the total hemispherical values of absorptivity and emissivity are also equal when $T_1 = T_2$. For bodies whose emissivities do not vary with wavelength (black or grey bodies), these overall values of absorptivity and emissivity are the same even when $T_1 \neq T_2$. But, for real surfaces, when $T_1 \neq T_2$, the deviation from this law increases as $T_1 - T_2$ increases. For example, in the tropics, white clothing has an advantage in that white has a low absorptivity for the high temperature, short wavelength, radiation from the Sun but a high emissivity for its own low-temperature radiation.

EXAMPLE 1.7

A sheet of glass is placed over a number of black objects in direct sunlight. The glass transmits the short wavelength radiation from the Sun but absorbs 90 per cent of the longer wavelength radiation from the objects themselves. Calculate the temperature of the objects for a glass temperature of 20°C (293 K).

Neglecting convection and conduction, the glass receives σT_b^4 from the objects and absorbs $0\cdot9\sigma T_b^4$. The glass itself radiates $0\cdot9\sigma T_g^4$ from each surface.

At equilibrium:

$$2 \times 0\cdot9\sigma T_g^4 = 0\cdot9\sigma T_b^4,$$

$$T_b = T_g \times 2^{1/4},$$

$$= 293 \times 1\cdot19,$$

$$= 348 \text{ K or } 75°C.$$

This value is not an accurate one but serves to illustrate the increased temperatures which can be obtained by shielding with glass, an effect (greenhouse effect) which relies on the variation of absorptivity of the glass with wavelength.

In many cases, however, it is sufficiently accurate to assume that materials are non-selective emitters, that is, they are grey and, therefore, obey Kirchhoff's and Lambert's laws.

1.15. BLACK-BODY ENCLOSURE

The near blackness of carbon blacks has already been referred to. It is often preferable to obtain an even closer approximation by means of an enclosure called the black-body enclosure. This is simply an enclosure maintained at a uniform temperature which has a small hole in it. Any radiation which enters the hole from outside is eventually absorbed after successive reflections inside, so that the absorptivity is effectively equal to 1. Kirchhoff's law may now be invoked to show that $\epsilon = 1$.

1.16. DATA ON EMISSIVITIES FOR REAL SURFACES

The emissivities of a number of surfaces are given in Table 1.3. The values are approximate since emissivity depends very much upon the conditions of the surface. Most non-metallic substances have high emissivities and are usually considered grey. Radiation from electrical conductors, especially polished metals, is markedly different. Emissivities are much lower and vary considerably with wavelength (Fig. 1.10);

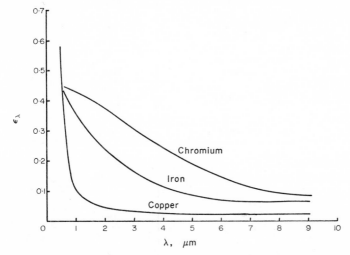

FIG. 1.10. Variations of emissivity with wavelength for some metals (Love, 1968).

radiation leaving these surfaces is not diffuse and reflected radiation may well be described by the ordinary laws of optics. Consequently, although the overall emissivities of metals are listed with those of other materials some care must be taken in their use. In general, dirt and oxidation considerably increase the emissivities of most surfaces, both dirt and oxides being poor conductors (Table 1.3).

Over small temperature ranges, emissivities do not vary greatly. When considering the absorption of radiation from a high-temperature source to a much cooler surface it is usually necessary to use an emissivity corresponding to the lower temperature and an absorptivity corresponding to the high temperature.

|TABLE 1.3. EMISSIVITY OF SOME COMMON
MATERIALS (Hottel and Sarofim, 1967)

Material	Emissivity
Aluminium, polished	0·095
oxidized	0·20
Brass, polished	0·03
dull plate	0·60
Copper, polished	0·02
oxidized	0·80
Iron, polished	0·20
oxidized	0·70
Steel, polished	0·07
oxidized	0·80
Stainless steels	0·2 to 0·7
Alumina	0·40
Asbestos	0·95
Brick, red	0·93
building	0·45
fireclay	0·75
Glass	0·95
Silica	0·4

Furthermore, the emissivity of many surfaces varies with the angle of emission. Few experimental data on the directional variation of emissivity are available; some workers (Kreith, 1967) assume $\epsilon/\epsilon_n = 1\cdot2$ for metallic surfaces and $\epsilon/\epsilon_n = 0\cdot96$ for non-metallic surfaces, where ϵ is the average emissivity throughout the hemispherical solid angle of 2π steradians and ϵ_n is the emissivity in the direction normal to the surface.

Even when the emissivity does vary with wavelength, an average emissivity or absorptivity for the wavelength band in which the bulk of the radiation is emitted or absorbed is often sufficiently accurate for calculation. If the distribution of the monochromatic emissivity is known, the average emissivity over the whole thermal spectrum can be obtained from

$$\epsilon = \frac{\int \epsilon_\lambda E_{b\lambda} d\lambda}{\sigma T^4}.$$

(1.19)

The total average absorptivity (a) for a surface can be obtained similarly by evaluating the fraction

$$a = \frac{\int a_\lambda E_{b\lambda} d\lambda}{\sigma T^4}$$

(1.20)

where $E_{b\lambda}$ and T are characteristic of the incident radiation.

EXAMPLE 1.8

The spectral emissivity of a surface at a temperature of 300 K varies with wavelength as shown in Fig. 1.11a. Calculate

(a) the average emissivity and
(b) the average absorptivity for black-body radiation at 1000 K.

The average emissivity is defined by eq. (1.19). The integral can be conveniently evaluated graphically. The function $E_{b\lambda}/\sigma T^4$ can be determined from Fig. 1.3 (a plot of $E_{b\lambda}/T^5$) and is plotted as a function of wavelength in Fig. 1.11b. The integral is obtained by plotting $\epsilon_\lambda E_{b\lambda}/_\delta T^4$ (Fig. 1.11b) and measuring the area under the curve to give $\epsilon = 0.0375$. Although ϵ_λ may be independent of temperature, the average value of the emissivity will be a function of temperature because the relative position of the ϵ_λ and $E_{b\lambda}/\sigma T^4$ curves change with temperature.

A similar procedure is used to evaluate the average absorptivity from eq. (1.20) but the temperature involved here is that of the incident radiation (1000 K). The functions $E_{b\lambda}/\sigma T^4$ and $\epsilon_\lambda E_{b\lambda}/\sigma T^4$ are plotted in Fig. 1.11c. The area under the curve gives an average absorptivity of 0.5.

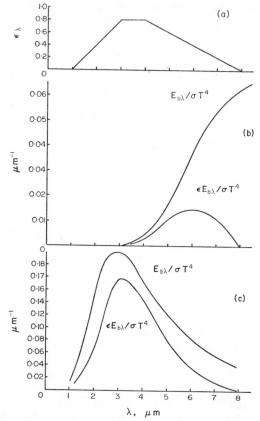

FIG. 1.11. Determination of average values of absorptivity and emissivity.

1.17. SOLAR RADIATION

The Sun is a body of hot gases surrounded by a layer of rarefied gas. When viewed from the Earth, the Sun appears as a flat disc called the photosphere. It is of the order of $1·39 \times 10^9$ m ($8·65 \times 10^5$ miles) in diameter and its mean distance from the Earth is $149·0 \times 10^9$ m ($92·6 \times 10^6$ miles). In many cases, the Sun can be treated as a point source because the mean solid angle of the Sun relative to an observer on the Earth is about a half a degree.

The amount of solar radiation incident upon a surface normal to that radiation and situated just outside the Earth's atmosphere is $1 \cdot 39$ kW/m² and is known as the solar constant.

EXAMPLE 1.9

If the amount of solar radiation incident upon the Earth corresponds to 64,110 kW/m² emitted from the surface of a black body, calculate its temperature.

$$q = \sigma T^4,$$

$$64,110 = 56 \cdot 7 \times (T/1000)^4,$$

$$T = 5799 \text{ K.}$$

The spectral distribution of solar radiation shows that the Sun does approximate to a black body and only departs from black-body behaviour in the ultra-violet region ($0 \cdot 1$ to $0 \cdot 4 \mu$m). Consequently, for engineering calculations, the Sun can be assumed to be a black body at a temperature of 5800 K. We have already calculated (Example 1.3) the proportion of this radiation in the visible and infra-red regions of the spectrum.

The amount of solar radiation reaching the Earth's surface is lower than the solar constant because of absorption and scattering by the Earth's atmosphere (approximately 145 km or 90 miles thick). It is convenient to express the amount of radiation incident on the surface of the Earth normal to the radiation from the Sun by

$$G_n = G_0 t_a{}^m \qquad (1.21)$$

where G_0 is the solar constant, m the relative air mass (ratio of the actual path length to the shortest possible path) and t_a is the transmission coefficient for unit mass of air.

The value of t_a depends on atmospheric conditions. A mean value of $0 \cdot 7$ is often used, with a value of $0 \cdot 81$ for a clear day and $0 \cdot 62$ for a cloudy one. The value of m depends on the zenith angle, z (referred to by astronomers as the zenith distance), which is the angle between a line connecting the Sun and the centre of the Earth and the normal to the surface of the Earth. On the assumption that the thickness of the atmosphere is negligible compared to the radius of the Earth, the

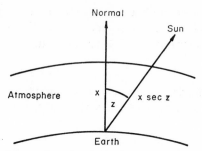

FIG. 1.12. Attenuation of radiation by the Earth's atmosphere.

relative path length of the atmosphere is sec(z) (Fig. 1.12), an approximation which is sufficiently accurate for $0 < z < 80°$. At angles greater than 80°, radiation is negligible.

For a surface not normal to the Sun's radiation, the incident radiation on the surface is

$$G_i = G_n \cos(i). \tag{1.22}$$

If the receiving surface is horizontal, $i = z$.

The zenith angle depends on the Sun's position in the sky relative to an observer on the surface. In turn, the Sun's position depends on the simultaneous motion of the Earth revolving in a plane (the ecliptic plane) once every 365·25 days around the Sun and spinning around its own axis once every 24 hours or at the rate of $\pi/(12 \times 3600)$ rad/s. The axis of rotation of the Earth is tilted at an angle of 23·5° with respect to the Sun's ecliptic axis.

Calculations of the position of the Sun require the adoption of some convenient system of coordinates such as those based on the following assumptions.

(1) The Earth is stationary.
(2) The heavenly bodies have been projected outward along lines which extend from the centre of the Earth to a sphere of infinite radius called the celestial sphere which possesses the same centre as that of the Earth.
(3) The celestial sphere rotates from east to west with respect to the Earth about the Earth's axis connecting the poles at a speed of 360° 59·14′ per 24 hours.

Thus, the geographical position of a celestial body is that point on the surface of the Earth which has the body in its zenith, that is, directly overhead. It is expressed in terms of latitude and longitude. The celestial equivalent of latitude is called the declination (d) and that of longitude is the Greenwich hour angle (GHA). The local hour angle of a body (h) is measured westward around the equator from the meridian of a given location, so that

$h = \text{GHA} - $ longitude west of Greenwich

or $h = \text{GHA} + $ longitude east of Greenwich.   (1.23)

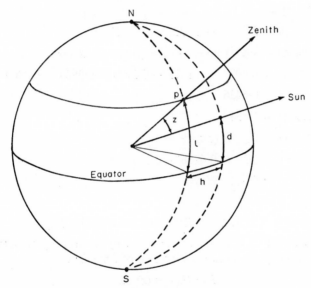

FIG. 1.13. Zenith, latitude, declination and hour angles.

Data on Greenwich hour angles and declinations may be obtained conveniently from a *Nautical Almanac*.

It can be shown (Fig. 1.13) that the zenith angle can be calculated from terrestrial coordinates by

$$\cos (z) = \sin (l) \sin (d) + \cos (l) \cos (d) \cos (h) \qquad (1.24)$$

where l is the latitude, d the declination and h the local hour angle.

EXAMPLE 1.10

Calculate the amount of solar radiation incident on Leeds, England (latitude N 53° 48′; longitude 1° 32′ west of Greenwich on 29 August 1973 at Greenwich Mean Time 14.00 hours). The solar constant is $1\cdot39$ kW/m² and the transmission coefficient of the atmosphere is $0\cdot7$.

From a 1973 *Nautical Almanac:*

Day	Hour	GHA	d
29.8.73	14	29° 47′	9° 17′
(Wednesday)			

From eq. (1.23)

$$h = 29° 47′ - 1° 32′ = 28° 15′$$

$$\cos(z) = \sin(l)\sin(d) + \cos(l)\cos(d)\cos(h)$$

$$= 0\cdot8070 \times 0\cdot1613 + 0\cdot5906 \times 0\cdot9869 \times 0\cdot8809$$

$$= 0\cdot6436$$

$$\sec(z) = 1\cdot554$$

$$G = G_0 \cos(z) t_a{}^{\sec(z)}$$

$$= 1\cdot39 \times 0\cdot6436 \times 0\cdot5745$$

$$= 0\cdot514 \text{ kW/m}^2.$$

1.17.1. *Solar Radiation received in a Day*

The amount of solar energy received by a surface just outside the Earth's atmosphere during an interval of time dt is

$$G = G_0 \cos(z)\, dt, \tag{1.25}$$

but
$$dt = dhdt/dh = dh/\Omega \tag{1.26}$$

where Ω is the angular velocity of the Earth $[\pi/(12 \times 3600) = \pi/43{,}200$ rad/s].

Consequently, the total energy received per unit area in a day is

$$G = 2 \times 4\cdot32 \times 10^4/\pi \int_0^H G_0 \cos(z)\, dh \tag{1.27}$$

where H is the hour angle between noon and sunrise or sunset.

For a particular location on the Earth, l is constant and, during one day, the declination is also effectively constant. Consequently, eq. (1.27) can be written as

$$G = 8.64 \times 10^4 \, G_0[H \sin (l) \sin (d) + \cos (l) \cos (d) \sin (H)]/\pi.$$
$$(1.28)$$

EXAMPLE 1.11

For the data in Example 1.10, determine the amount of solar energy received during the day.

By interpolation from the tables for sunrise and sunset in the *Nautical Almanac* for 29.8.73:

Sunrise at latitude N 53° 48: 0457 hours,
Sunset   : 1849 hours.

$$H = \pi(6 \text{ hr } 56 \text{ min})/12 = 1.815 \text{ rad} = 104°.$$

From eq. (1.28):

$$G = 8.64 \times 10^4 \times 1.39[1.815 \times 0.8070 \times 0.1613$$
$$+ 0.5906 \times 0.9869 \times 0.9703]/\pi$$

$$= 3.07 \times 10^4 \text{ kJ/m}^2 \text{ (neglecting attenuation by the Earth's atmosphere).}$$

CHAPTER 2

Direct Radiative Transfer

THE direct exchange of radiation between two surfaces depends on two factors:

(1) the view which the surfaces have of each other and
(2) the emitting and absorbing characteristics of the surfaces.

In some cases, however, it is possible to solve a problem of direct radiative exchange by making simple assumptions about these factors.

2.1. SIMPLIFIED SITUATIONS

2.1.1. *Black Body completely surrounded by Black Surfaces*

The body, at a temperature T_1, emits radiation at a rate of $\sigma T_1{}^4$ and receives radiation at the rate of $\sigma T_2{}^4$ from the surrounding surfaces at a temperature T_2. Thus, the net heat transferred to the body is

$$q = \sigma(T_1{}^4 - T_2{}^4). \tag{2.1}$$

2.1.2. *Grey Body completely surrounded by Black Surfaces*

If the body is grey with emissivity ϵ, it will emit $\epsilon \sigma T_1{}^4$ and receive $\sigma T_2{}^4$ but will only absorb $\epsilon \sigma T_2{}^4$ (since absorptivity equals emissivity according to Kirchhoff's law). The net heat flow is

$$q = \epsilon \sigma(T_1{}^4 - T_2{}^4). \tag{2.2}$$

2.1.3. *Small Grey Body completely surrounded by Grey Surfaces*

If the surroundings are grey, eq. (2.2) is no longer valid because some of the radiation emitted by the body is reflected back to it. But the

26

expression may still be used without significant loss of precision if the body is small compared to the surrounding surfaces because the body intercepts only a negligible proportion of the reflected radiation.

EXAMPLE 2.1

A hot steel billet, $3 \times 1 \times 1$ m, has 95 per cent of its surface exposed. The billet has an emissivity (grey) of 0·30, its thermal conductivity is very high and the surroundings are at 30°C (303 K). The specific heat and density of the steel are 500 J/kg K and 7800 kg/m³, respectively. Assuming that negligible heat is lost by conduction through the supports and that the convective heat losses are small, calculate the time for the billet to cool from 1000° to 800°C (1273 to 1073 K).

Since the thermal conductivity is very high, the rate of cooling is determined by the rate of radiative heat transfer from the surface and is not limited by the rate at which heat is transferred from the interior of the material to the surface. Furthermore, since the dimensions of the billet are much smaller than those of the surrounding surfaces, these may be treated as black.

From eq. (2.2) the net rate of heat transfer from the billet is

$$Q = A\epsilon\sigma(T^4 - T_s^4)$$

where A is the surface area of the billet and T and T_s are the temperatures of the billet and surrounding surfaces respectively.

Q may be expressed as the rate of decrease of the enthalpy of the billet, that is, dH/dt.

Thus
$$dH/dt = d(\rho VcT)/dt$$
$$= \rho Vc dT/dt$$

and
$$-\rho Vc dT/dt = A\epsilon\sigma(T^4 - T_s^4).$$

Since $T \gg T_s$:
$$T^4 - T_s^4 = T^4$$

and
$$-\rho Vc dT/dt = A\epsilon\sigma T^4$$

or
$$\int_0^t dt = -\frac{\rho Vc}{A\epsilon\sigma} \int_{1273}^{1073} dT/T^4.$$

Integrating

$$t = \frac{\rho V c}{A \epsilon \sigma 3} \left[\frac{1}{1073^3} - \frac{1}{1273^3} \right]$$

$$= \frac{7800 \times 3 \times 500}{0{\cdot}95 \times 14 \times 0{\cdot}30 \times 56{\cdot}7 \times 10^{-3} \times 3} \left[\frac{1}{1{\cdot}073^3} - \frac{1}{1{\cdot}273^3} \right]$$

$$= 5597 \text{ s} = 1{\cdot}55 \text{ hr.}$$

EXAMPLE 2.2

A thermocouple is used to measure the temperature of air flowing in a large duct having a wall temperature of 700 K. The thermocouple indicates a temperature of 900 K. Calculate the true temperature of the air if the convective heat-transfer coefficient is known to be 0·16 kW/m² K under these conditions. The emissivity of the thermocouple wire is 0·2.

At thermal equilibrium, the net loss of heat from the thermocouple to the walls by radiation equals the net gain of heat from the air by convection (assuming that transfer of heat by other mechanisms is negligible).

Thus

$$A \epsilon \sigma (T_t{}^4 - T_w{}^4) = A k (T_g - T_t)$$

where T_t, T_w and T_g are the temperatures of the thermocouple, wall and gas, respectively, A the area of the thermocouple bead and k the convective heat transfer coefficient:

$$\epsilon \sigma (T_t{}^4 - T_w{}^4) = k (T_g - T_t)$$

and
$$T_g = \epsilon \sigma (T_t{}^4 - T_w{}^4)/k + T_t$$

$$= 29{\cdot}5 + 900$$

$$= 929{\cdot}5 \text{ K.}$$

Apart from these simple geometries, the view which the surfaces have of each other has to be determined for each geometry.

2.2. VIEW FACTORS AND EXCHANGE AREAS

The view factor from one surface to another is defined as the fraction of the total radiation emitted by the one surface which is directly incident on the other. Thus, from Fig. 2.1, the radiation leaving surface 1 and directly incident upon surface 2 is

$$Q = (\epsilon_1 A_1 E_{b1}) \times F_{12} \qquad (2.3)$$

where F_{12} is the view factor from surface 1 to 2.

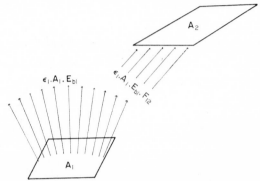

Fig. 2.1. Direct transmission of radiation from one surface to another.

As a consequence of the definition of view factor as a fraction, the summation of all the view factors from a surface to all other surfaces receiving radiation from it is 1.

For an enclosure comprising n surfaces:

$$F_{11} + F_{12} + F_{13} + \ldots + F_{1n} = 1 \qquad (2.4)$$

or

$$\sum_{j=1} F_{1j} = 1. \qquad (2.4a)$$

The term F_{11} is included to take into account radiation from surface 1 which is directly incident upon itself, for example, when surface 1 is concave.

From eq. (1.12) the radiation emitted from a small area of a black surface (ΔA) at an angle ϕ and transmitted through a solid angle Δw is

$$\Delta Q = I_n \Delta A \cos \phi \Delta w \qquad (1.12)$$

For a surface of emissivity ϵ_1, eq. (1.12) may be written as

$$\Delta Q = \epsilon_1 E_b \Delta A \cos \phi \Delta w / \pi. \tag{2.5}$$

Expressing this equation in differential form (illustrated by Fig. 2.2) allows the total radiation transmitted directly from a surface A_1 to a surface A_2 to be obtained from

$$Q = \int_w \int_{A_1} \epsilon_1 E_b dA_1 \cos \phi_1 dw / \pi \tag{2.6}$$

$$= \epsilon_1 E_b \int_w \int_{A_1} dA_1 \cos \phi_1 dw / \pi. \tag{2.6a}$$

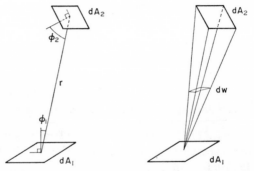

Fig. 2.2. View factor geometry for two surfaces of differential area.

Equation (2.6a) implies that (a) ϵ_1 is independent of the angle of emission and position over the whole area and (b) the temperature over the whole area is uniform.

By comparison of eq. (2.6) with eq. (2.3):

$$F_{12} = \frac{1}{A_1} \int_w \int_{A_1} \frac{dA_1 \cos \phi_1 dw}{\pi}. \tag{2.7}$$

The area viewed from surface 1 is $dA_2 \cos \phi_2$ and

$$dw = dA_2 \cos \phi_2 / r^2. \tag{2.8}$$

Substituting eq. (2.8) in eq. (2.7):

$$F_{12} = \frac{1}{A_1} \int_{A_1} \int_{A_2} \frac{dA_1 \cos \phi_1 \, dA_2 \cos \phi_2}{\pi r^2}. \qquad (2.9)$$

From the symmetry of the integral equation [eq. (2.9)]:

$$A_1 F_{12} = A_2 F_{21}. \qquad (2.10)$$

The product $A_1 F_{12}$ (or $\overline{12}$) is called the exchange area between surfaces 1 and 2 so that

$$\overline{12} = \overline{21}. \qquad (2.11)$$

It is useful to express eq. (2.4) in terms of exchange areas:

$$\sum_{j=1}^{n} \overline{1j} = A_1. \qquad (2.12)$$

EXAMPLE 2.3

A furnace can be considered as a rectangular box (5 m long, 4 m wide and 3 m high) comprising three surfaces of constant temperature, the ceiling (1), walls (2) and floor (3). If $F_{12} = 0.67$, calculate the other view factors for this geometry.

$$F_{13} = 1 - F_{12} = 0.33.$$

From symmetry: $F_{31} = F_{13}$

and $F_{32} = F_{12},$

$$A_1 F_{12} = A_2 F_{21},$$

$$F_{21} = 20 \times 0.67/54 = 0.248.$$

From symmetry: $F_{23} = F_{21}$

$$F_{22} = 1 - F_{23} - F_{21}$$

$$= 1 - 2F_{21}$$

$$= 0.504.$$

2.2.1. *Evaluation of View Factors*

The evaluation of eq. (2.9) is often difficult and, in any case, can generally be avoided by using view factors which have already been derived. Values for a variety of geometries are available in the literature either as relatively simple functions or in chart form. For example, for parallel, directly opposed, plane circular discs of radius r_1 and r_2 and separated by distance d:

$$F_{12} = [x - (x^2 - 4r_2{}^2/r_1{}^2)^{1/2}]/2 \qquad (2.13)$$

where

$$x = 1 + d^2/r_1{}^2 + r_2{}^2/r_1{}^2.$$

2.2.1.1. *Charts*

Other data on view factors have been presented in chart form and Leuenberger and Person (1956), Kreith (1962), Sparrow and Cess (1966) and Hottel and Sarofim (1967) are useful sources. Figure 2.3 is an example of this kind of presentation; it shows view factors between parallel plates and touching perpendicular plates.

2.2.1.2. *Derivations from known geometries*

In some cases, view factors for certain geometries may be deduced from the known view factors of other geometries. For example, consider the case of two perpendicular plates shown in Fig. 2.4.

The view factors from surfaces $(A + B)$ to C and from B to C (touching perpendicular plates) may be determined from a chart such as that shown in Fig. 2.3 but the view factor from surface A to C can also be obtained by simple algebra.

The radiation from B to C plus the radiation from A to C equals the radiation from $(A + B)$ to C, or, expressed in terms of the exchange areas,

$$\overline{BC} + \overline{AC} = \overline{(A + B)C} \qquad (2.14)$$

and

$$\overline{AC} = \overline{(A + B)C} - \overline{BC}. \qquad (2.15)$$

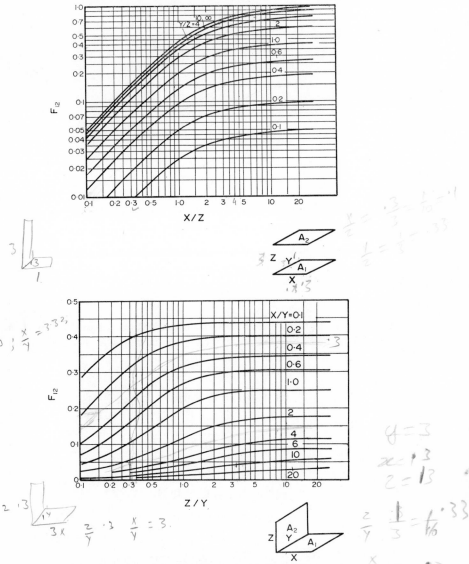

FIG. 2.3. (*Above*) View factors for identical, parallel rectangular plates.
(*Below*) View factors for touching perpendicular plates.

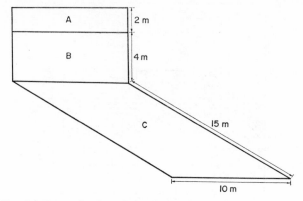

FIG. 2.4. Determination of view factors for perpendicular plates.

EXAMPLE 2.4

Calculate the view factor between areas A and C in Fig. 2.4.

From eq. (2.15):

$$\overline{AC} = \overline{(A + B)C} - \overline{BC}.$$

The exchange areas $\overline{(A + B)C}$ and $\overline{BC}$ are obtained from the chart in Fig. 2.3 using $X/Y = 0.6$ and 0.4 ($X = 6$ and 4 m and $Y = 10$ m) and $Z/Y = 1.5$ ($Z = 15$ m).

Thus

$$\overline{(A + B)C} = 6 \times 10 \times F_{(A+B)C}$$

$$= 60 \times 0.28$$

$$= 16.8 \text{ m}^2;$$

$$\overline{BC} = 4 \times 10 \times F_{BC}$$

$$= 40 \times 0.33$$

$$= 13.2 \text{ m}^2.$$

Therefore

$$\overline{AC} = 16.8 - 13.2 = 3.6 \text{ m}^2$$

and

$$F_{AC} = 3.6/20 = 0.18.$$

Other deductions of view factors or exchange areas can be made in cases where symmetry or similarity exists between the case in question and a known geometry. For example, in Fig. 2.5 the exchange area $\overline{12}$ equals the exchange area $\overline{34}$ because, for every pair of elements dA_1 and dA_2, there exists a similar pair dA_3 and dA_4 separated by the same distance r and orientated identically (Yamauti principle).

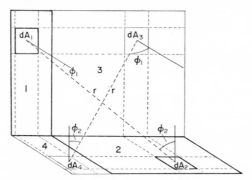

FIG. 2.5. Application of Yamauti principle.

2.2.1.3. Crossed-string method

The crossed-string method of determining exchange areas $(A_1 F_{12})$ can be used when the surfaces have

(a) lengths much greater than their widths and separation,
(b) constant cross-sections normal to their lengths and
(c) constant separation along their lengths.

If the two surfaces $(A_1 + A_2)$ in Fig. 2.6a are represented by their cross-sections shown in Fig. 2.6b, it can be shown (Hottel and Sarofim, 1967) that the exchange area per unit length is

$$\overline{12} = \frac{L_1 + L_2 - L_3 - L_4}{2} \tag{2.16}$$

where L_1, L_2, L_3 and L_4 are the lengths of the lines produced by stretching strings tightly between the extreme points of the surfaces. Equation

(2.16) is usually expressed as

$$\overline{12} = \frac{\text{sum of the crossed strings} - \text{sum of the uncrossed strings}}{2}.$$

(2.17)

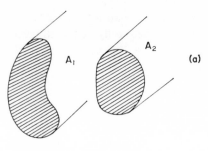

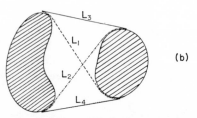

FIG. 2.6. The crossed-string method.

EXAMPLE 2.5

Calculate the view factor between two parallel tubes 6 m long, 60 mm diameter and separated (centre to centre) by 100 mm.

Since the tubes are parallel, of uniform cross-section and their lengths are much greater than their diameters and separations, the crossed-string method is appropriate (Fig. 2.7).

From symmetry, $L_1 = L_2$ and $L_3 = L_4$.

Hence $\overline{12} = (2L_1 - 2L_3)/2 = L_1 - L_3,$

$L_3 = 100$ mm,

$L_1 = AC = AE + EG + GC$

$= 2(AE + EO).$

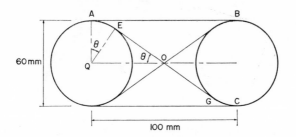

FIG. 2.7. The crossed-string method used to determine exchange areas between two long parallel circular tubes.

But
$$EO = (QO^2 - QE^2)^{1/2}$$
$$= (50^2 - 30^2)^{1/2}$$
$$= 40 \text{ mm},$$

and
$$AE = QE \times \theta$$
$$= QE \times \sin^{-1}(QE/QO)$$
$$= 30 \times 0\cdot644$$
$$= 19\cdot3 \text{ mm}.$$

Therefore
$$L_1 = 118\cdot6 \text{ mm}$$

and
$$\overline{12} = 18\cdot6 \text{ mm}.$$

Hence
$$F_{12} = 18\cdot6/60\pi$$
$$= 0\cdot099.$$

If one of the surfaces is infinitesimally narrow (Fig. 2.8a), it may be shown that the view factor from dA to surface A_2 is

$$F_{dA\,2} = (\sin \psi_2 + \sin \psi_1)/2. \tag{2.18}$$

In Fig. 2.8b, the view factor is

$$F_{dA\,2} = (\sin \psi_2 - \sin \psi_1)/2. \tag{2.19}$$

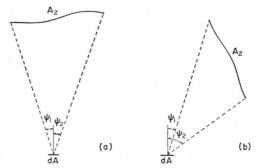

FIG. 2.8. The crossed-string method used to determine the view factors for two surfaces, one of which is infinitesimally small.

EXAMPLE 2.6

An extensive flame, resulting from the combustion of a spillage of liquid fuel, is contained behind a wall. If the effective black-body temperature of the flame is 900°C (1173 K) and the wall remains effectively cold, at what distance from the wall is the radiative flux a maximum and what is its value? The heights of the flame and wall are 11 m and 1 m respectively.

The radiant energy incident on a narrow strip of ground of width Δx, length l, at a distance x from the flame front is

$$Q = HlF_{f\Delta x}\sigma T_f^4$$
$$= l\Delta x F_{\Delta x f}\sigma T_f^4$$

where T_f is the flame temperature and $F_{\Delta x f}$ is the view factor from the strip to the flame (Fig. 2.9).

The flux, $q = Q/\Delta x l$

$$= F_{\Delta x f}\sigma T_f^4$$

and is a maximum where $dF/dx = 0$.

$F_{\Delta x f}$ is given by eq. (2.18):

$$\sin \psi_2 = x/(x^2 + h^2)^{1/2},$$
$$\sin \psi_1 = x(x^2 + (H + h)^2)^{1/2},$$

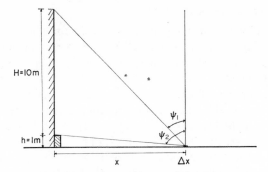

FIG. 2.9. Radiation from an extensive flame.

$$F_{\Delta xf} = \frac{1}{2}\left[\frac{x}{(x^2 + h^2)^{1/2}} - \frac{x}{(x^2 + (H + h^2)^{1/2}}\right],$$

$$\frac{dF}{dx} = \frac{1}{2}\left[\frac{1}{(x^2 + h^2)^{1/2}} - \frac{x^2}{(x^2 + h^2)^{3/2}}\right.$$

$$\left. - \frac{1}{(x^2 + (H + h)^2)^{1/2}} + \frac{x^2}{(x^2 + (H + h)^2)^{3/2}}\right]$$

$$= \frac{1}{2}\left[\frac{h^2}{(x^2 + h^2)^{3/2}} - \frac{(H + h)^2}{(x^2 + (H + h)^2)^{3/2}}\right].$$

The point of maximum flux occurs where

$$h^2(x^2 + (H + h)^2)^{3/2} = (H + h)^2(x^2 + h^2)^{3/2}.$$

Raising both sides of the equation to the power of $2/3$:

$$h^{4/3}(x^2 + (H + h)^2) = (H + h)^{4/3}(x^2 + h^2),$$

$$x^2 + 11^2 = 11^{4/3} x^2 + 11^{4/3},$$

$$x = 2 \cdot 03 \text{ m}.$$

$F_{\Delta xf}$ at this point is $0 \cdot 358$ and the flux is

$$q = 0 \cdot 358 \times 56 \cdot 7 \times (1 \cdot 173)^4,$$

$$= 38 \cdot 4 \text{ kW/m}^2.$$

2.2.1.4. *Numerical methods*

Equation (2.9) may be integrated analytically for a limited number of geometries. These include the parallel plate, perpendicular plate and cylindrical geometries already mentioned. For more complex geometries, numerical techniques have to be used, and this generally requires the use of a computer. Techniques for numerical integration are well established but, in this case, the integration is rather complex.

Alternatively the Monte Carlo method may be used. In this method, radiation is represented by a large sample of beams which are emitted in directions dictated by the selection of a series of random numbers according to the following procedure. (It has been shown that the radiation can be represented satisfactorily by this technique by Howell (1968).) Each beam emitted by a surface is tested to see if it intercepts the second surface and the view factor (F_{12}) is given by the ratio of the number of intercepted beams to the total number emitted. The following sequence of operations describes the procedure:

1. Define the two surfaces of interest geometrically (using cartesian, cylindrical or spherical coordinates as appropriate).
2. Select a point at random on surface 1.
3. Select a direction at random for a beam of radiation emitted from this point. This direction can be defined by the circumferential angle θ and the cone angle ϕ. The two angles are obtained from random numbers (R_1 and R_2) in the range 0 to 1 by

$$\theta = 2\pi R_1 \qquad (2.20)$$

and
$$\cos \phi = R_2{}^{1/2}. \qquad (2.21)$$

The derivation of eqs. (2.20) and (2.21) has been explained by Howell (1968).

4. Determine whether the emitted beam intercepts the second surface and count it as either intercepted or not.
5. Go back to step 1 and repeat this procedure a large number of times, the number depending on the precision required.
6. The view factor is given by the ratio of the number of beams intercepted to the total number emitted.

In practice, a computer is required to carry out the large number of operations.

EXAMPLE 2.7

Determine the view factor between two parallel, directly opposed, plane circular discs of different diameter by means of the Monte Carlo technique.

R_1, R_2 and R_3 are random numbers in the range 0 to 1.

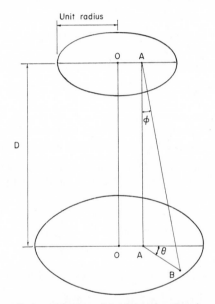

FIG. 2.10. Monte Carlo method used to determine the view factor for two parallel, directly opposed, plain circular discs.

A point A is selected at random on the surface of the first disc (Fig. 2.10). Due to symmetry, A is simply described by the distance from the centre of the disc OA and

$$OA^2 = R_1.$$

The square of the distance from the centre is used because a point is chosen at random from the area of the disc and not from the diameter.

The direction of a beam of radiation from point A is defined by the angles θ and ϕ (Fig. 2.10) according to the procedure described in Appendix 2.

$$\cos \phi = R_2^{1/2}$$

and
$$\theta = 2\pi R_3.$$

The point at which this beam strikes the plane in which the second disc is situated is given by B in Fig. 2.10. The distance O^*B can be determined from

$$O^*B^2 = O^*A^2 + A^*B^2 + 2O^*AO^*B \cos \theta$$

$$= R_1 + D^2 \tan^2\phi + 2R_1^{1/2}D \tan \phi \cos \theta$$

where $\tan^2\theta = 1/R_2 - 1$.

If O^*B is equal to or less than the radius of the second disc, then the beam is counted as a "hit".

The whole procedure, starting with the selection of the point A, is then repeated and the ratio of the number of "hits" to the total number of beams selected gives the view factor from disc 1 to disc 2.

2.3. TOTAL ENERGY TRANSFER AMONG BLACK SURFACES

2.3.1. *Summation*

The net heat exchange between two surfaces i and j is

$$Q_{ij} = A_i F_{ij} E_{bi} - A_j F_{ji} E_{bj}. \tag{2.22}$$

Because $A_i F_{ij} = A_j F_{ji}$ [eq. (2.10)], eq. (2.22) may be expressed as

$$Q_{ij} = A_i F_{ij}(E_{bi} - E_{bj}). \tag{2.23}$$

The net heat exchange between a surface i and the rest of the surfaces within an enclosure comprising n black surfaces is

$$Q_i = \sum_{j=1}^{n} A_i F_{ij}(E_{bi} - E_{bj}). \tag{2.24}$$

2.3.2. *Networks*

An inspection of eq. (2.23) shows that an analogy may be drawn between radiative heat transfer and the flow of electric current through a resistor. Thus, voltage is the analogue of black-body emissive power (E_b), current of net heat exchange (Q) and electrical resistance of reciprocal exchange area ($R_{ij} = 1/A_iF_{ij}$). Figure 2.11 shows the analogy.

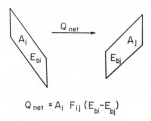

$$Q_{net} = A_i\, F_{ij}\, (E_{bi} - E_{bj})$$

$$I = \frac{1}{R}.(V_i - V_j)$$

FIG. 2.11. Electrical analogy for direct radiative transfer between black surfaces.

If there are more than two surfaces in the system, a network of resistors may be constructed, the value of each resistor being set equal to the appropriate value of reciprocal exchange area. When appropriate voltages are applied to the nodes, the currents, and thus net heat exchange, can be determined. For example, Fig. 2.12 represents an enclosure comprising four black surfaces (Oppenheim, 1956).

When heat is transferred by radiation (H) and convection (q_{conv}) to the surface of a body, the temperature of the surface rises until the following equilibrium is established:

$$H + q_{conv} = q_{cond} + \epsilon E_b + \rho H$$

and
$$H = (q_{cond} - q_{conv}) + \epsilon E_b + \rho H \qquad (2.25)$$

where q_{cond} is the heat conducted from the surface through the body.

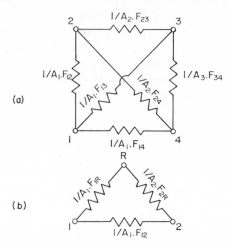

FIG. 2.12. (a) Network representation of an enclosure comprising four black surfaces.

(b) Network representation of an enclosure comprising two black surfaces and one re-radiating surface.

In the case of a hot refractory wall in a furnace, $(q_{cond} - q_{conv})$ is small and H is approximately equal to $\epsilon E_b + \rho H$. In this situation, the surface re-radiates almost all the radiant energy incident upon it and is called a re-radiator.

In the network representation, the refractory surface appears as a node to which a potential is not initially assigned. Figure 12b shows this for an enclosure comprising two black surfaces and a refractory surface.

EXAMPLE 2.8

For the system represented by Fig. 12b, determine the temperature to which the refractory surface will rise.

The potential at node R is given by

$$(E_{bR} - E_{b1})/(E_{b2} - E_{b1}) = R_{1R}/(R_{1R} + R_{2R})$$

$$= \frac{1/A_1 F_{1R}}{1/A_1 F_{1R} + 1/A_2 F_{2R}}$$

where R_{1R} and R_{2R} denote the resistances to flow between surface 1 and the refractory and between surface 2 and the refractory respectively.

$$E_{bR} = \frac{A_1 F_{1R} E_{b1} + A_2 F_{2R} E_{b2}}{A_1 F_{1R} + A_2 F_{2R}}.$$

For $A_1 = A_2$:

$$E_{bR} = \frac{F_{1R} E_{b1} + F_{2R} E_{b2}}{F_{1R} + F_{2R}}.$$

Setting $E_b = \sigma T^4$

$$T_R = \left[\frac{F_{1R} T_1{}^4 + F_{2R} T_2{}^4}{F_{1R} + F_{2R}} \right]^{1/4}.$$

CHAPTER 3

Total Exchange of Radiation within an Enclosure containing a Non-absorbing Medium

IN A real situation, we are concerned not with transfer of radiation between two black surfaces but with transfer among more than two non-black surfaces, transfer which involves successive reflections and absorptions at all the surfaces. Together, these surfaces make up an enclosure and the simplest analysis involves determining the net heat transfer to one of these surfaces all of which are assumed to be grey. The analysis of heat transfer to grey, instead of black, surfaces is conveniently done by using the concept of radiosity. The use of this concept is not confined to a grey surface but is always appropriate when reflection of radiation from a surface takes place.

3.1. LEAVING FLUX OR RADIOSITY

Consider the fluxes incident upon and leaving a grey surface (Fig. 3.1). H is the incident flux, ρH that which is reflected, ϵE_b that which is

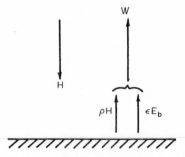

FIG. 3.1. Radiant energy incident upon and leaving a surface.

46

emitted by the surface by virtue of its temperature and W the total flux leaving the surface.

$$W = \epsilon E_b + \rho H. \tag{3.1}$$

Since the leaving flux (W) or radiosity incorporates the radiation reflected from a surface, any analysis based on radiosity automatically takes reflected radiation into account.

The net radiant flux (the net loss of energy per unit time per unit area) can be expressed either by

$$q = W - H = (\epsilon E_b + \rho H) - H = \epsilon(E_b - H), \tag{3.2}$$

or $$q = W - H = W - (W - \epsilon E_b)/\rho = \epsilon(E_b - W)/\rho. \tag{3.3}$$

3.2. RADIATIVE HEAT TRANSFER IN AN EMPTY ENCLOSURE

Consider an enclosure comprising n surfaces at temperatures T_1, $T_2, T_3, \ldots, T_n$ (Fig. 3.2). From eq. (3.1), we can write for each surface i:

$$W_i = \epsilon_i E_{bi} + \rho_i H_i. \tag{3.4}$$

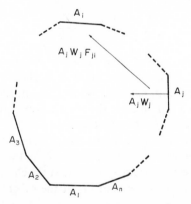

FIG. 3.2. Enclosure comprising n surfaces.

An amount of radiation ($A_j W_j$) leaves every surface of area A_j and a fraction of this (F_{ji}) is directly incident upon surface i. Thus, the total energy incident upon surface i is the sum of terms like $A_j W_j F_{ji}$ [eq. (3.5)].

(We do not have to consider reflected radiation because it is incorporated in the radiosity term.)

$$A_i H_i = W_1 A_1 F_{1i} + W_2 A_2 F_{2i} + \ldots + W_n A_n F_{ni}. \tag{3.5}$$

Since $\quad A_i F_{ij} = A_j F_{ji},$

$$A_i H_i = W_1 A_i F_{i1} + W_2 A_i F_{i2} + \ldots + W_n A_i F_{in}. \tag{3.6}$$

Or $\qquad H_i = \sum_{j=1}^{n} W_j F_{ij}. \tag{3.7}$

Substituting eq. (3.7) into eq. (3.4) leads to

$$W_i = \epsilon_i E_{bi} + \rho_i \left(\sum_{j=1}^{n} W_j F_{ij} \right) \tag{3.8}$$

where $i = 1, 2, 3, \ldots, n$.

An equation of this kind is produced for each surface and provides n linear simultaneous equations containing n unknown values of the leaving fluxes. After solution of these equations, the net heat transferred from each surface is provided by eq. (3.3) multiplied by the area of the surface, that is,

$$Q_i = \frac{\epsilon_i (E_{bi} - W_i) A_i}{(1 - \epsilon_i)}. \tag{3.9}$$

Problems involving two- or three-surface enclosures need only involve hand calculations; enclosures of more than three surfaces generally require some computational aid. W and E_b have to be evaluated accurately since eq. (3.3) involves the subtraction of these values which may be of the same order of magnitude.

It is important to note that the surfaces referred to may not correspond to a single plane surface like, for example, the roof of a furnace. The surface examined in the analysis may, at the choice of the designer, consist of a small part of the roof or, at the other extreme, the roof and all the walls.

EXAMPLE 3.1

A Dewar flask for storing liquid nitrogen (boiling point 126 K) is constructed of concentric cylinders, the inner one 0·5 m long, 0·2 m in diameter and the outer one 0·52 m long and 0·22 m in diameter. The neck of the flask is very narrow and the surfaces enclosing the vacuum are coated with aluminium of emissivity 0·04 (assumed grey). If the outer cylinder is at a temperature of 21°C, calculate the net radiative heat transfer to the nitrogen.

Since all the radiation from the inner surface (1) is incident on the outer surface (2), the view factor (F_{12}) from surface 1 to surface 2 is unity. Hence

$$F_{21} = A_1/A_2$$

$$= \frac{0·5 \times \pi \times 0·2 + 2\pi \times 0·01}{0·52 \times \pi \times 0·22 + 2\pi \times 0·012}$$

$$= 0·866.$$

And
$$F_{22} = 1 - F_{21}$$

$$= 0·134,$$

$$\epsilon_1 = \epsilon_2 = 0·04,$$

$$\rho_1 = \rho_2 = 0·96,$$

$$E_{b1} = 56·7 \times (0·126)^4 = 0·143 \text{ kW/m}^2,$$

$$E_{b2} = 56·7 \times (0·294)^4 = 0·424 \text{ kW/m}^2.$$

The radiosity equations are

$$W_1 = \epsilon_1 E_{b1} + \rho_1 W_2 F_{12},$$

$$W_2 = \epsilon_2 E_{b2} + \rho_2 (W_1 F_{21} + W_2 F_{22}).$$

Solution of these equations gives

$$W_1 = 0·229 \text{ kW/m}^2,$$

$$W_2 = 0·238 \text{ kW/m}^2.$$

And the net heat transfer to the nitrogen is

$$Q = A_1 \epsilon_1 (W_1 - E_{b1})/\rho_1$$

$$= 3·37 \text{ W}.$$

3.3. REFRACTORY WALLS

The assumption that refractory walls in a furnace act as re-radiators (Section 2.3.2) simplifies the calculations of radiative heat transfer within the furnace enclosure. Each re-radiating surface has an effective reflectivity of 1 and an effective emissivity of 0.

EXAMPLE 3.2

A muffle furnace is considered to be a rectangular box (Fig. 3.3) in which the ceiling acts as the heating surface with emissivity of 0·92. The walls and floor are refractory with an emissivity of 0·69. Calculate the net rate of heat transfer to the floor of the furnace on the assumption that the refractory side walls effectively re-radiate all the heat incident on them. The ceiling and floor temperatures are 1200 and 600 K respectively.

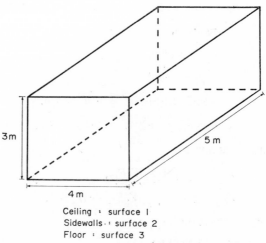

Ceiling : surface 1
Sidewalls : surface 2
Floor : surface 3

FIG. 3.3. Furnace enclosure comprising three isothermal surfaces.

The ceiling, side walls and floor of the furnace are considered as three surfaces 1, 2 and 3 respectively (Fig. 3.3). Since the refractory side walls are assumed to re-radiate all the radiation incident upon them, the emissivity of the surface is effectively zero and the radiosity equations may be written as

$$W_1 = \epsilon_1 E_{b1} + \rho_1(W_2 F_{12} + W_3 F_{13}),$$
$$W_2 = W_1 F_{21} + W_2 F_{22} + W_3 F_{23},$$
$$W_3 = \epsilon_3 E_{b3} + \rho_3(W_1 F_{31} + W_2 F_{32})$$

or

$$W_1 - \rho_1 F_{12} W_2 - \rho_1 F_{13} W_3 = \epsilon_1 E_{b1}$$
$$-F_{21} W_1 + (1 - F_{22})W_2 - F_{23} W_3 = 0$$
$$-\rho_3 F_{31} W_1 - \rho_3 F_{32} W_2 + W_3 = \epsilon_3 E_{b3}.$$

F_{13} $(=F_{31})$ can be obtained from a standard chart (Fig. 2.3) and the other view factors in the following way:

$$F_{13} = F_{31} = 0 \cdot 33,$$
$$F_{12} = F_{32} = 1 - F_{13} = 0 \cdot 67,$$
$$F_{21} = F_{23} = F_{12} A_1 / A_2 = 0 \cdot 248,$$
$$1 - F_{22} = 2 \times F_{23} = 0 \cdot 496.$$

Coupled with the values of the view factors, the following data are also substituted in the radiosity equations:

Surface	1	2	3
Emissivity	0·92	0	0·69
Reflectivity	0·08	1	0·31
Area m²	20	54	20
$\epsilon E_b = \epsilon \sigma T^4$ kW/m²	108	0	5·07

The equations can now be solved for W_3:

$$W_3 = \frac{\begin{vmatrix} 1 \cdot 0 & -0 \cdot 054 & 108 \\ -0 \cdot 248 & 0 \cdot 496 & 0 \\ -0 \cdot 102 & -0 \cdot 208 & 5 \cdot 07 \end{vmatrix}}{\begin{vmatrix} 1 \cdot 0 & -0 \cdot 054 & -0 \cdot 026 \\ -0 \cdot 248 & 0 \cdot 496 & -0 \cdot 248 \\ -0 \cdot 102 & -0 \cdot 208 & 1 \cdot 0 \end{vmatrix}}$$

$$= 31 \cdot 57 \text{ kW/m}^2.$$

The net heat transfer to the floor is

$$Q = A_3\epsilon_3(W_3 - E_{b3})/\rho_3$$
$$= 1078 \text{ kW.}$$

3.4 ENCLOSURE COMPRISING GREY AND BLACK SURFACES

When one of the surfaces in the enclosure can be assumed to be black, the analysis must be modified to determine the energy incident upon the surface (H) and not the leaving flux (radiosity, W).

EXAMPLE 3.3

Using the data in Example 3.2, calculate the net heat transfer to the floor of the furnace assuming it to be black.

The net heat transfer to the floor is

$$Q = A_3H_3 - A_3E_{b3}$$
$$= A_1W_1F_{13} + A_2W_2F_{23} - A_3E_{b3}.$$

The radiosity equations are

$$W_1 = \epsilon_1E_{b1} + \rho_1W_2F_{12} + \rho_1E_{b3}F_{13},$$
$$W_2 = W_1F_{21} + W_2F_{22} + E_{b3}F_{23}.$$

These may be solved to give

$$W_1 = 111 \cdot 0 \text{ kW/m}^2$$
and
$$W_2 = 59 \cdot 2 \text{ kW/m}^2$$
and
$$Q = 1378 \text{ kW.}$$

3.5. CONDUCTION OF HEAT THROUGH FURNACE WALLS

In the examples considered so far, we have used values of the surface temperatures, a procedure compatible with making measurements of those temperatures. However, similar calculations can be made without a knowledge of these temperatures and such calculations make use of the ambient temperature and the thermal conductivities of the materials comprising the walls.

EXAMPLE 3.4

The stock completely covers the floor of a furnace which is of the type described in Example 3.2. The stock and floor combined is 0·3 m thick and possesses an effective thermal conductivity of 0·196 kW/m K. If the ambient temperature is 23°C, calculate the surface temperature of the stock which has an emissivity of 0·69. Assume that the outer skin temperature of the floor is the same as ambient.

At equilibrium, the net heat transfer to the floor by radiation equals that transferred by conduction through the stock and floor.

The net heat transfer to the stock by radiation is

$$Q_r = A_3\epsilon_3(W_3 - E_{b3})/\rho_3.$$

W_3 may be expressed as the ratio of determinants as in Example 3.2:

$$W_3 = \frac{\begin{vmatrix} 1\cdot0 & -0\cdot054 & 108 \\ -0\cdot248 & +0\cdot496 & 0 \\ -0\cdot102 & -0\cdot208 & 0\cdot69E_{b3} \end{vmatrix}}{\begin{vmatrix} 1\cdot0 & -0\cdot054 & -0\cdot026 \\ -0\cdot248 & 0\cdot496 & -0\cdot248 \\ -0\cdot102 & -0\cdot208 & 1\cdot0 \end{vmatrix}}$$

$$= (0\cdot333E_{b3} + 11\cdot02)/0\cdot427$$

$$= 0\cdot780E_{b3} + 25\cdot8$$

and $$Q_r = A_3 \times 0\cdot69 \times (0\cdot78E_{b3} + 25\cdot8 - E_{b3})/0\cdot31$$

$$= A_3 \times (57\cdot4 - 0\cdot49E_{b3})$$

$$= A_3 \times (57\cdot4 - 0\cdot49 \times 56\cdot7 \times (T_3/1000)^4).$$

The net heat transfer by conduction is

$$Q_c = A_3 \times 0\cdot196 \times (T - 296)$$

$$= A_3 \times 196 \times (T/1000 - 0\cdot296).$$

But $Q_r = Q_c$, that is,

$$57\cdot4 - 27\cdot8 \times (T/1000)^4 = 196 \times (T/1000 - 0\cdot296)$$

$$(T/1000)^4 = 4\cdot15 - 7\cdot04 \times (T/1000).$$

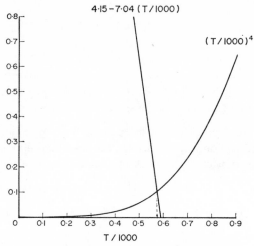

FIG. 3.4. Graphical solution for Example 3.4.

Both sides of this equation are plotted in Fig. 3.4 as a function of $T/1000$ and the surface temperature of the stock determined from the point of intersection to be 575 K.

3.6. ZONING

When there are significant temperature variations throughout an enclosure, it becomes necessary to subdivide (conceptually) the enclosure into a large number (n) of surface zones, each zone being characterized by the near constancy of its temperature and emissivity. The application of eq. (3.8) produces n simultaneous equations which can be solved for the values of radiosity corresponding to each surface. But n^2 view factors have to be obtained or determined for n surfaces. Since view factor data in the literature are limited, the problem can often be solved by making use of a simplifying procedure, that of subdividing the enclosure by a number of imaginary planes which are then used in the analysis. For example, Fig. 3.5 shows a furnace of rectangular cross-section subdivided into sections by three planes, each section comprising six surfaces (including the imaginary plane). The analysis now consists of two steps:

1. Equations like eq. (3.8) are written for each section.
2. The equations for each section are linked by the concept that the radiosity (W) from an imaginary plane in one section equals the incident flux (H) on that plane from an adjacent section (Fig. 3.6). Thus, $W_{1A} = H_{1B}$ and $W_{1B} = H_{1A}$.

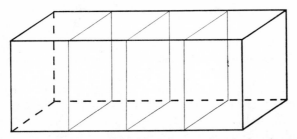

FIG. 3.5. Furnace enclosure divided into three sections by imaginary planes.

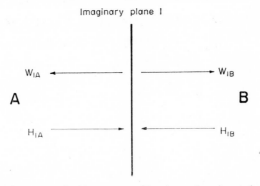

FIG. 3.6. Radiant energy incident upon and leaving an imaginary plane surface.

EXAMPLE 3.5

The fall in temperature along a cylindrical furnace, 7 m long and 1 m in diameter, may be approximated by a temperature of 1100°C (1373 K) along the first 2 m, 800°C (1073 K) for the next 3 m and 650°C (923 K) for the final 2 m. The end faces are at the temperatures of their respective sections and all surfaces have emissivities of 0·75. Calculate the net heat transfer to the cold end section.

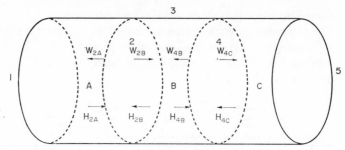

FIG. 3.7. Cylindrical enclosure divided into three sections by imaginary planes.

The furnace is divided into three sections A, B and C by the two imaginary planes 2 and 4 (Fig. 3.7). The surfaces in the three sections, A, B and C, are labelled 1, 3 and 5 respectively.

The radiosity equations for the three real surfaces are

$$W_1 = \epsilon E_{b1} + \rho(W_1 F_{11} + W_{2A} F_{12}),$$

$$W_3 = \epsilon E_{b3} + \rho(W_{2B} F_{32} + W_3 F_{33} + W_{4B} F_{34}),$$

$$W_5 = \epsilon E_{b5} + \rho(W_{4C} F_{54} + W_5 F_{55}).$$

The linking equations for surface 2 are

$$W_{2A} = H_{2B},$$

$$W_{2B} = H_{2A}.$$

Since

$$A_2 H_{2B} = A_3 W_3 F_{32} + A_4 W_{4B} F_{42}$$

and

$$A_2 H_{2A} = A_1 W_1 F_{12}$$

then

$$W_{2A} = W_3 F_{23} + W_{4B} F_{24},$$

$$W_{2B} = W_1 F_{21}.$$

Similarly, for surface 4

$$W_{4C} = H_{4B},$$

$$W_{4B} = H_{4C}.$$

Since

$$A_4 H_{4B} = A_2 W_{2B} F_{24} + A_3 W_3 F_{34},$$

$$A_4 H_{4C} = A_5 W_5 F_{54},$$

then

$$W_{4C} = W_{2B} F_{42} + W_3 F_{43},$$

$$W_{4B} = W_5 F_{45}.$$

Substituting the linking equations in the radiosity equations:

$$W_1 = \epsilon E_{b1} + \rho(W_1 F_{11} + W_3 F_{23} F_{12} + W_5 F_{45} F_{12} F_{24}),$$

$$W_3 = \epsilon E_{b3} + \rho(W_1 F_{21} F_{32} + W_3 F_{33} + W_5 F_{45} F_{34}),$$

$$W_5 = \epsilon E_{b5} + \rho(W_1 F_{21} F_{42} F_{54} + W_3 F_{43} F_{54} + W_5 F_{55}).$$

The view factors are

$$F_{24} = F_{42} = 0.04,$$

$$F_{23} = F_{43} = 1 - 0.04 = 0.96,$$

$$F_{32} = F_{23} A_2 / A_3 = 0.08 = F_{34},$$

$$F_{33} = 1 - F_{32} - F_{34} = 0.84,$$

$$F_{21} = F_{45} = 1.0,$$

$$F_{12} = F_{54} = A_2 / A_1 = 0.111,$$

$$F_{11} = F_{55} = 1 - F_{12} = 0.889.$$

The three radiosity equations are solved to give

$$W_5 = 42.6 \text{ kW/m}^2$$

and the net heat transfer to surface 5 is

$$Q = A_5 \epsilon (W_5 - E_{b5}) / \rho$$

$$= 31.2 \text{ kW.}$$

3.7. NETWORK METHOD

The network method referred to in Section 2.3.2 can also be used for reflecting surfaces. Equation (2.23) gives the direct heat exchange between two black surfaces. Similarly, the direct heat exchange between two reflecting surfaces i and j is

$$Q_{ij} = A_i F_{ij}(W_i - W_j). \tag{3.10}$$

This equation shows that a network may be constructed for radiative heat transfer among reflecting surfaces by applying potentials corresponding to radiosity at the nodes. In order to apply the appropriate value of radiosity to each node, use is made of eq. (3.3) which gives the net flow of heat from the surface:

$$Q_{net} = A\epsilon(E_b - W)/\rho. \tag{3.3}$$

The net flow of heat from the surface is represented in the network by a current flowing through a resistance $\rho/A\epsilon$ between two potentials corresponding to E_b and W. Figure 3.8 shows the representation of an enclosure comprising three reflecting surfaces.

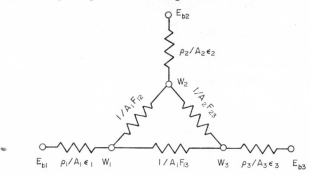

FIG. 3.8. Network representation of an enclosure comprising three surfaces.

EXAMPLE 3.6

Using the data in Example 3.2, calculate the net rate of heat transfer to the floor of the furnace by means of the network method.

Figure 3.9 shows the network representation of radiative heat transfer within the furnace.

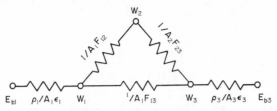

FIG. 3.9. Network representation of an enclosure comprising two radiating and one re-radiating surface.

The resistance to flow between potentials E_{b1} and E_{b3} is

$$\frac{\rho_1}{A_1\epsilon_1} + \frac{1/A_1F_{12} + 1/A_3F_{32}}{1 + A_1F_{13}(1/A_1F_{12} + 1/A_3F_{32})} + \frac{\rho_3}{A_3\epsilon_3}$$

$$= \frac{0\cdot08}{0\cdot92 \times 20} + \frac{2/13\cdot4}{1 + 6\cdot6 \times 2/13\cdot4} + \frac{0\cdot31}{0\cdot69 \times 20}$$

$$= 0\cdot102$$

$$Q_{\text{net}} = (E_{b1} - E_{b3})/0\cdot102$$

$$= 56\cdot7 \times (1\cdot2^4 - 0\cdot6^4)/0\cdot102$$

$$= 1080 \text{ kW.}$$

CHAPTER 4

Radiative Heat Transfer within an Enclosure containing an Absorbing Medium

4.1. EMISSION AND ABSORPTION OF RADIATION BY AN ABSORBING MEDIUM

The attenuation of a parallel beam of radiation on passage through a uniformly absorbing medium of thickness L is

$$I = I_0 e^{-KL} \tag{4.1}$$

where I_0 is the intensity of radiation incident upon the medium, I the intensity of the transmitted radiation and K the attenuation or extinction coefficient.

If the radiation is not parallel, the various beams will travel different distances through the medium and will be attenuated by different amounts. The overall attenuation can still be described by eq. (4.1) if an appropriate average length (the mean beam length, L_m) is chosen. In general, the mean beam length is a function of the geometry and the attenuation coefficient.

The fraction of radiation transmitted (transmissivity) by the medium is

$$\tau = e^{-KL_m}. \tag{4.2}$$

If the radiation not transmitted is assumed to be absorbed, then

$$a = 1 - e^{-KL_m}. \tag{4.3}$$

For a grey medium:

$$\epsilon = a = 1 - e^{-KL_m}. \tag{4.4}$$

Consequently, the radiation emitted by the medium, characterized by the length L_m, can be expressed as

$$q = \epsilon\sigma T^4$$

$$= (1 - e^{-KL_m})\sigma T^4. \tag{4.5}$$

4.1.1. *Determination of Mean Beam Lengths*

For the case of radiation transmitted from the surface of a hemispherical envelope enclosing an absorbing medium to a very small area at the centre of its base, the path length for each beam of radiation arriving at the area is constant and equal to the radius (R). Consequently, the mean beam length is equal to the radius and

$$\tau = e^{-KR} \tag{4.6}$$

and

$$\epsilon = 1 - e^{-KR}. \tag{4.7}$$

For other geometries, it may be possible to obtain an analytical expression for the attenuation and hence the emissivity although it will not be of this simple form. For example, Booth (1949) has shown the emissivity of a spherical medium to its enclosing surface to be

$$\epsilon = 1 - 2(1 - (KD - 1)e^{-KD})/K^2 D^2 \tag{4.8}$$

where D is the diameter of the sphere.

This equation can be written in the form

$$\epsilon = 1 - e^{-fKD} \tag{4.9}$$

where f is a function of K and D.

Comparison with eq. (4.4) shows that the mean beam length is fD. In general, the mean beam length for any shape of absorbing medium can be expressed as

$$L_m = fD \tag{4.10}$$

where D is a characterizing dimension of the enclosure. For a sphere, D is the diameter; for a cube, it is the length of one edge; and for a parallelopiped, D is usually the shortest edge.

Although f is a function of KD (Fig. 4.1), a constant value is used to evaluate the emissivity in practice. Hottel and Sarofim (1967) have tabulated two values of f, one (f_0) evaluated at $KD = 0$ and the second (f_m) chosen to minimize the consequent error in ϵ over the practical range of values of KD. In addition, it was assumed that the medium was grey, an assumption which is maintained throughout this chapter. Table 4.1 lists values of f for a number of geometries.

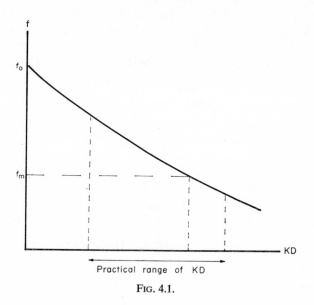

FIG. 4.1.

If radiation from an absorbing medium to its entire bounding surface is being considered, L has been found to be approximately equal to $3 \cdot 5V/A$ where V refers to the volume of the medium and A to the area of its bounding surface.

EXAMPLE 4.1

A grey gas at a temperature of 800°C has an attenuation coefficient of $0 \cdot 15$ m^{-1} (reciprocal metres) and is contained in an enclosure of side lengths 1 m, 1 m and 4 m. Using the tabulated mean beam lengths in

TABLE 4.1 MEAN BEAM LENGTHS (ABSORBING MEDIUM TO A SURFACE)

Shape	Characterizing dimension	f_0 at $KD = 0$	f_m
Right circular cylinders	Diameter		
1. Height = diameter, radiating to			
whole surface		0·6	0·6
2. Height = 0·5 × diameter, radiating to			
(a) end		0·47	0·43
(b) side		0·52	0·46
(c) whole surface		0·50	0·45
3. Height = 2 × diameter, radiating to			
(a) end		0·73	0·60
(b) side		0·82	0·76
(c) whole surface		0·80	0·73
Rectangular parallelopipeds	Shortest edge		
1. Cube, radiating to side		0·67	0·60
2. 1 × 1 × 4 radiating to			
(a) 1 × 4 face		0·90	0·82
(b) 1 × 1 face		0·86	0·71
(c) all faces		0·89	0·81
3. 1 × 2 × 6 radiating to			
(a) 2 × 6 face		1·18	—
(b) 1 × 6 face		1·24	—
(c) 1 × 2 face		1·18	—
(d) all faces		1·20	—

Table 4.1, calculate the energy incident on a 1 × 1-m face and a 1 × 4-m face and, hence, determine the total energy incident on all faces. Compare this value with that obtained by using a single value of the mean beam length for the whole of the enclosing surface and by using a value of $3·5V/A$.

1 m × 4 m face: $L_m = 0·82$ m,

$$Q = 4·0 \times (1 - e^{-0·15 \times 0·82}) \times 56·7 \times 1·073^4$$

$$= 34·9 \text{ kW}.$$

1 m × 1 m face: $L_m = 0·71$ m,

$$Q = 7·59 \text{ kW}.$$

The total energy incident on all the surfaces is

$$Q = 4 \times 34{\cdot}9 + 2 \times 7{\cdot}59$$
$$= 155 \text{ kW.}$$

For the enclosure as a whole,

$$L_m = 0{\cdot}81 \text{ m}$$

and $$Q = 18 \times (1 - e^{-0{\cdot}81 \times 0{\cdot}15}) \times 56{\cdot}7 \times 1{\cdot}073^4$$
$$= 156 \text{ kW.}$$

Using a value of $L_m = 3.5V/A = 0{\cdot}78$ m

$$Q = 149 \text{ kW.}$$

Perhaps the most common absorbing medium that we have to deal with is a suspension of solid or liquid particulate material in a gas such as the Earth's atmosphere or a special atmosphere such as a furnace atmosphere. In any case, the analysis is basically the same; furnace atmospheres are used as a basis for discussion in what follows immediately and attenuation by the Earth's atmosphere is considered in Chapter 5.

The particulate and gaseous components of flames or combustion products in a furnace both contribute to their ability to attenuate radiation and, therefore, their absorptivity and emissivity.

4.1.2. *Particles in Combustion Products*

The luminosity of luminous gas flames is caused by solid particles which arise from the thermal decomposition of vapour-phase hydrocarbons in the flame. They are composed of carbon and very heavy hydrocarbons and are 0·005 to 0·15 μm in size. These particles are referred to as soot (black smoke).

The asphaltenes (mainly polycyclic aromatic hydrocarbons with paraffin side chains) in heavy fuel oils form coke when the oil is sprayed into the furnace. Consequently, there are present in a flame of this kind porous coke particles of about the same size as the original oil droplets. In addition, soot particles arising from the incomplete combustion of the volatile material are also present.

Pulverized fuel flames contain particles varying in size from 200 μm downwards. The size of these particles is related to the size of the coal fed to the furnace. In this country, approximately 85 per cent by weight of the coal fed to pulverized fuel burners in power station boilers is less than 75 μm in size. When the coal enters the furnace some swelling may take place and the size begins to diminish after the volatile material has been burned. The composition of the particles varies from high carbon contents to ash.

Radiation incident on a suspension of particles in a gas is partly transmitted, partly scattered and partly absorbed. Scattering can take place by a number of mechanisms, namely,

1. diffraction at the particle/gas interface,
2. refraction at the gas/particle interface and re-emission from the particles after internal reflection and further refraction and
3. reflection from the particle surface.

These mechanisms occur, to a greater or lesser extent, with particles of all sizes and a complete solution to the problem of determining the extent of each mechanism involves the use of Maxwell's equations for the interaction of electromagnetic waves with spherical particles. Mie, for example, has treated the general case of light propagation in a turbid medium. The Maxwell analysis involves the parameter $\pi d/\lambda$ (d is the particle diameter) and the complex optical property $n(1 - ik')$ where n is the refractive index, k' the absorption index of the particle material and $i^2 = -1$. The computation is normally considered too difficult and simpler estimates of the contribution of the particulate material to the emissivity of the flame can be made for particles within certain size ranges.

1. *Large particles* ($d > 5\lambda/\pi \simeq 2\lambda$)

For large black particles, present in low concentrations (fractional volume of space occupied by particles $\ll 1$), the fractional decrease in the intensity of radiation as it passes through the suspension is proportional to the area intercepted by the particles (A)

$$A = c_n A_n \qquad (4.11)$$

where c_n is the number of particles per unit volume of the suspension and A_n the projected area per particle in the direction of radiation.

Alternative expressions for the area of obscuration involve the volume and mass concentrations. They are

$$A = c_v A_v \qquad (4.12)$$

and

$$A = c_m A_m \qquad (4\cdot13)$$

where c_v is the fractional volume of particles in the suspension, A_v the projected area of the particles per unit volume of the particle material, c_m the mass concentration of particles (kg/m^3) and A_m the projected area per unit mass of particle material.

It can be shown (Herdan, 1960) that

$$A_v = S_v/4 \qquad (4.14)$$

where S_v is the surface area of the particles per unit volume of the particle material. In this case, eq. (4.1) becomes

$$I = I_0 e^{-c_v S_v L/4} \qquad (4\cdot15)$$

and $K = c_v S_v/4$ in eqs. (4.2), (4.3) and (4.4).

The surface area of the particles per unit volume of combustion products $(c_v S_v)$ can be calculated from values of the mean diameter of the particles, the mass of fuel and air $(m_f$ and $m_a)$ and the densities of fuel and the combustion products $(\rho_f$ and $\rho_p)$ by means of eqs. (4.16) to (4.19).

1. Surface area per unit volume of fuel (S_v):

Volume of a spherical particle of fuel $= \pi d^3/6$ m^3. \qquad (4.16)

Surface area of a particle \qquad $= \pi d^2$ \quad m^2. \qquad (4.17)

Dividing these values: \qquad $S_v = 6/d$ \quad m^2/m^3. \qquad (4.18)

2. Surface area per unit volume of combustion products $(c_v S_v)$:

$$c_v S_v = \frac{6 m_f \rho_p}{d(m_f + m_a)\rho_f}. \qquad (4.19)$$

EXAMPLE 4.2

A heavy fuel oil of density 965 kg/m^3 is atomized to produce a spray characterized by a surface-mean droplet diameter of 100 μm. The oil is burned with 20 per cent excess air, the resulting coke residues being of the same size as the atomized droplets. If the combustion products are at a temperature of 1500°C (1773 K) in a cylindrical combustion chamber, 1 m in diameter and 2 m long, calculate the emissivity of the combustion products. Assume that the mean molecular weight of the combustion products is 29·2 and 13·6 kg of air is required for the stoichiometric combustion of 1 kg of fuel.

From the perfect gas law:

$$\rho_p = MP/RT$$

where ρ_p, M, P and T are the density, molecular weight, total pressure and temperature of the combustion products respectively.

For a total pressure of 1 bar

$$\rho_p = 29·2/(0·08314 \times 1773) = 0·198 \text{ kg/m}^3,$$

$$\rho_p/\rho_f = 0·198/965 = 2·05 \times 10^{-4},$$

$$(m_f + m_a)/m_f = (1 + 13·6 \times 1·2)/1 = 17·32,$$

$$c_v S_v = \frac{6 \times 2·05 \times 10^{-4}}{10^{-4} \times 17·32} = 0·710 \text{ m}^{-1},$$

$$\epsilon = 1 - e^{-0·71 L_m/4}.$$

For $\qquad L_m = 0·73, \epsilon = 0·121.$

2. *Small particles* ($d < 0·6\lambda/\pi \simeq 0·2\lambda$)

Soot particles in flames are less than 0·5 μm in size, that is, smaller than the wavelengths important in thermal radiation from a flame. Consequently, they cannot be considered to obscure and absorb radiation as shown by simple geometric optics. It is usual to calculate the hypothetical cross-sectional areas, relative to the areas of black particles of the same size, which are equivalent to attenuation (F_{at}) and scattering (F_{sc}) of radiation by the particles upon which the radiation falls. The relative absorption cross-sectional area is

$$F_{ab} = F_{at} - F_{sc}. \qquad (4.20)$$

An approximate solution of the Mie equation by Hawksley (1952) showed that scattering by particles of the size of soot particles is negligible and, consequently, the absorptivity is obtained from values of the attenuation. The attenuation of radiation by particles of this size is

$$F_{at} = 24\pi d f(n, k')/\lambda = F_{ab} \tag{4.21}$$

where $f(n, k')$ is a function of the refractive index (n) and the absorption index (k') of the particle material.

Since F_{ab} is proportional to $1/\lambda$ (ignoring the variation of n and k' with wavelength), a soot particle is markedly non-grey and the analysis must involve, initially, the monochromatic absorption coefficient.

For large particles, K was shown to be equal to $c_v S_v/4$ and, for a large black spherical particle, $K = 6c_v/4d = 3c_v/2d$.

For spherical soot particles

$$K_\lambda = 3F_{ab}c_v/2d$$

$$= 36\pi c_v f(n, k')/\lambda \tag{4.22}$$

$$= 36\pi c_m f(n, k')/\rho\lambda \tag{4.23}$$

where c_m is the concentration by mass of the particles in the suspension and ρ the density of the particles.

If the value of K_λ in eq. (4.22) is used to provide an estimate of the monochromatic emissivity, then

$$\epsilon_\lambda = 1 - \exp(-36\pi c_v f(n, k')L_m/\lambda)$$

$$= 1 - e^{-\kappa c_v L_m/\lambda} \tag{4.24}$$

where $\kappa = 36\pi f(n, k')$.

The average emissivity over the whole thermal spectrum may now be obtained from

$$\epsilon = \frac{\int_0^\infty \epsilon_\lambda E_{b\lambda} d\lambda}{\int_0^\infty E_{b\lambda} d\lambda}. \tag{4.25}$$

Substituting the Wien equation, $E_{b\lambda} = c_1\lambda^{-5}/e^{c_2/\lambda T}$, which is an approximation of the Planck expression valid for small values of λT, in eq. (4.25) gives

$$\epsilon = \frac{\int_0^\infty (1 - e^{-\kappa c_v L_m/\lambda})c_1\lambda^{-5}e^{-c_2/\lambda T}d\lambda}{\int_0^\infty c_1\lambda^{-5}e^{-c_2/\lambda T}d\lambda}$$

$$= \frac{1 - \int_0^\infty \exp(-(\kappa T c_v L_m + c_2)/\lambda T)\lambda^{-5}d\lambda}{\int_0^\infty \lambda^{-5}e^{-c_2/\lambda T}d\lambda}. \qquad (4.26)$$

Substituting $x = c_2/\lambda T$,

$$dx/d\lambda = -c_2/\lambda^2 T,$$

$$y = (c_2 + \kappa T c_v L_m)/\lambda T$$

and $\quad dy/d\lambda = -(c_2 + \kappa T c_v L_m)/\lambda^2 T$ in eq. (4.26) and eliminating λ gives

$$\epsilon = \frac{1 - T^4/(c_2 + \kappa c_v L_m T)^4 \int_0^\infty y^3 e^{-y}dy}{(T/c_2)^4 \int_0^\infty x^3 e^{-x}dx}.$$

The gamma functions $\left(\int_0^\infty y^3 e^{-y}dy \text{ and } \int_0^\infty x^3 e^{-x}dx \right)$ cancel to provide a value of the emissivity of

$$\epsilon = 1 - 1/(1 + \kappa c_v L_m T/c_2)^4. \qquad (4.27)$$

McCartney and Ergun (1958) showed that κ varied from 7·5 to 3·7 for coal flames when the ratio of hydrogen atoms to carbon atoms in the coal varied from 0 to 0·4. For oil flames, Field *et al.* (1967) deduced a value of 6·3 from the experimental data of Thring (1962).

EXAMPLE 4.3

Calculate the concentration of soot required to produce an emissivity of 0·2 for an oil flame at a temperature of 2000 K which is characterized by a mean beam length of 1 m. The density of the soot is 2000 kg/m³ and κ for the soot in the flame is 7·2.

$$\epsilon = 1 - 1/(1 + \kappa c_v L_m T/c_2)^4,$$

$$1 - 0·2 = 1/(1 + 7·2 \times c_v \times 1 \times 2000/1·439 \times 10^{-2})^4,$$

$$c_v = 5·7 \times 10^{-8}$$

and $c_m = 1·14 \times 10^{-4} \text{ kg/m}^3.$

4.1.3. Gases in Combustion Products

For gases, eq. (4.1) is written in the form

$$I = I_0 e^{-kpL} \qquad (4.28)$$

where p is the partial pressure of the absorbing gas.

Carbon dioxide and water are the important absorbing and emitting gases in combustion products although other gases (for example, methane) also absorb significant amounts of radiation. None of these gases, however, absorbs (or emits) over the whole of the thermal spectrum but only within bands in the spectrum. But data on emissivities are presented as if absorption does occur over the whole thermal range. Hottel (McAdams, 1954) has presented these data in graphical form (Figs. 4.2 to 4.6) and the overall emissivity of the combustion products due to carbon dioxide and water is obtained from

$$\epsilon_g = \epsilon_{CO_2} + \epsilon_{H_2O} - \Delta\epsilon$$
$$= \epsilon^*_{CO_2} C_{CO_2} + \epsilon^*_{H_2O} C_{H_2O} - \Delta\epsilon \qquad (4.29)$$

where $\epsilon^*_{CO_2}$ = emissivity of carbon dioxide at a total pressure of 1 bar and in the limit as the partial pressure of carbon dioxide approaches zero;

C_{CO_2} = correction for total pressure different from 1 bar and partial pressure of carbon dioxide different from zero;

$\epsilon^*_{H_2O}$ = emissivity of water for a total pressure of 1 bar and in the limit as the partial pressure of water approaches zero;

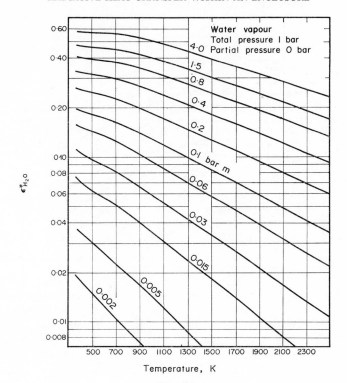

FIG. 4.2.

C_{H_2O} = correction for total pressure different from 1 bar and partial pressure of water different from zero;

$\Delta\epsilon$ = correction for the overlapping of the carbon dioxide and water bands.

Similarly,

$$a_g = a_{CO_2} + a_{H_2O} - \Delta a. \tag{4.30}$$

But

$$a_{CO_2} = a^*_{CO_2}(T_g/T_w)^{0.65} C_{CO_2}$$

and

$$a_{H_2O} = a^*_{H_2O}(T_g/T_w)^{0.45} C_{H_2O}.$$

Δa is evaluated at the temperature of the wall (T_w).

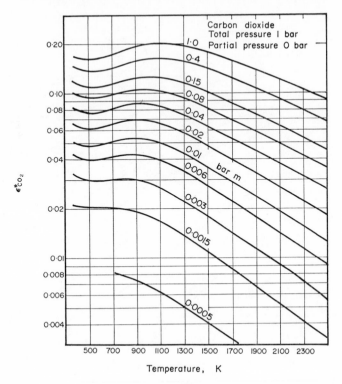

Fɪɢ. 4.3.

More recently, Leckner (1972) has presented data, obtained from spectroscopic measurements, for the emissivities of atmospheres containing carbon dioxide and water. These data differ somewhat from the data presented by Hottel, and Leckner has discussed the significance of the discrepancies.

Eхамрle 4.4

Calculate the heat radiated from a cylindrical mass of combustion products which are characterized by the following data:

Temperature: 1200 K.
Total pressure: 1 bar.

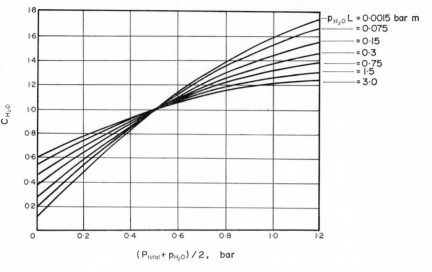

FIG. 4.4.

Partial pressures of carbon dioxide and water: 0·08 and 0·06 bar, respectively.

Diameter of combustion chamber: 1 m.

Length of combustion chamber: 2 m.

The mean beam length for a cylinder of length equal to twice the diameter is 0·73 × diameter (Table 4.1).

$$L_m = 0.73 \text{ m},$$

$$p_{CO_2} L_m = 0.08 \times 0.73 = 0.0584 \text{ m bar},$$

$$p_{H_2O} L_m = 0.06 \times 0.73 = 0.0438 \text{ m bar}.$$

Substituting values obtained from Figs. 4.2 to 4.6 in eq. (4.29) gives

$$\epsilon = 0.09 \times 1.0 + 0.066 \times 1.05 - 0.007$$

$$= 0.152.$$

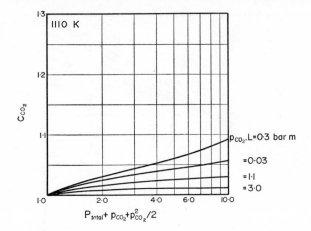

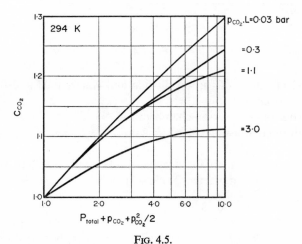

FIG. 4.5.

The radiant flux emitted by the combustion product is

$$q = \epsilon E_b$$
$$= 0{\cdot}152 \times 56{\cdot}7 \times 1{\cdot}2^4$$
$$= 17{\cdot}9 \text{ kW/m}^2.$$

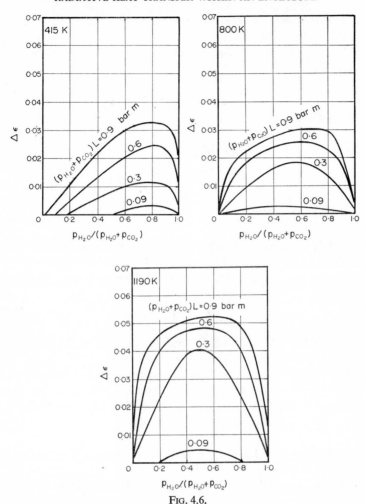

FIG. 4.6.

4.2. EFFECT OF AN ABSORBING MEDIUM ON THE RADIATIVE HEAT TRANSFER WITHIN AN ENCLOSURE

If an absorbing medium is present in the enclosure described by Fig. 3.2, eq. (3.4) is still valid but

1. radiation passing from one surface to another is attenuated by the medium and
2. by virtue of its temperature, the medium itself radiates energy to each surface,

$$W_i = \epsilon_i E_{bi} + \rho_i H_i. \tag{3.4}$$

Equation (3.7) must be rewritten to incorporate these factors

$$H_i = \sum_{j=1}^{n} W_j F_{ij}, \tag{3.7}$$

$$H_i = W_1 F_{i1} \tau_{i1} + W_2 F_{i2} \tau_{i2} + \dots + W_n F_{in} \tau_{in} + M_i \tag{4.31}$$

$$= \sum_{j=1}^{n} W_j F_{ij} \tau_{ij} + M_i \tag{4.32}$$

where τ_{ij} is the transmissivity of the medium between surfaces i and j and M_i the flux emitted by the medium which is incident upon surface i.

M_i can be written as $\epsilon_m E_{bm}$ and eq. (4.32) can be expressed as

$$H_i = \sum_{j=1}^{n} W_j F_{ij} \tau_{ij} + \epsilon_m E_{bm} \tag{4.33}$$

where ϵ_m is the emissivity of the medium and E_{bm} is the black-body emissive power of the medium.

Finally, n linear simultaneous equations can again be obtained which provide values for the radiosity of each surface of the enclosure. Each equation is a variant of eq. (3.8), incorporating the characteristics of the medium in the enclosure:

$$W_i = \epsilon_i E_{bi} + \rho_i \sum_{=1}^{n} W_j F_{ij} \tau_{ij} + \rho_i \epsilon_m E_{bm}. \tag{4.34}$$

The emissivity of the medium in eq. (4.34) is defined by

$$\epsilon_m = 1 - e^{-KL_{ms}} \tag{4.35}$$

where L_{ms} is the mean beam length evaluated by considering the radiation from the medium to the surface (such as those given in Table 4.1).

The transmissivity (τ_{ij}) in eq. (4.34) is

$$\tau_{ij} = e^{-KL_{ij}} \tag{4.36}$$

where L_{ij} must be determined by considering radiation from surface i which is transmitted to surface j. (Note that $L_{ij} = L_{ji}$ and $\tau_{ij} = \tau_{ji}$.)

Sometimes a single value of L_{ij} is assumed for all pairs of surfaces and eq. (4.34) can be written as

$$W_i = \epsilon_i E_{bi} + \rho_i \tau \sum_{j=1}^{n} W_j F_{ij} + \rho_i \epsilon_m E_{bm}. \qquad (4.37)$$

However, greater accuracy requires unique values of the transmissivity for each pair of surfaces. Dunkle (1964) has listed mean beam lengths for parallel and perpendicular plate geometries but data for a large number of geometries and values of K are not available.

EXAMPLE 4.5

The combustion products in a fuel-fired furnace, 6 m long and 2 m wide, can be assumed to be isothermal at a temperature of 1300 K and emit and absorb as a grey gas with an absorption coefficient of $0\cdot2$ m^{-1}. The walls and roof are at a temperature of 1100 K and have an emissivity of $0\cdot6$. The stock on the floor of the furnace has a surface temperature of 1000 K and an emissivity of $0\cdot7$, the distance between the surface of the stock and the roof being 1 m. Calculate the net heat transfer to the stock given the following data on mean beam lengths.

In this analysis, the roof and walls are treated as a single surface (1), the stock referred to as surface 2 and the combustion products as g.

$L_{g2} = 1\cdot18$ m (from Table 4.1),
$L_{g1} = 1\cdot20$ m (from Table 4.1 on the assumption that radiation is to all the surfaces),

$L_{11} = 5\cdot8$ m $\Big\}$ derived from Dunkle's data using the procedure
$L_{21} = 1\cdot15$ m $\Big\}$ outlined in Appendix 3.

The net heat transfer to the stock is

$$Q = A_2 \epsilon_2 (W_2 - E_{b2})/\rho_2$$

and the radiosity equations are

$$W_1 = \epsilon_1 E_{b1} + \rho_1 (W_1 F_{11} \tau_{11} + W_2 F_{12} \tau_{12} + \epsilon_{g1} E_{bg}),$$
$$W_2 = \epsilon_2 E_{b2} + \rho_2 (W_1 F_{21} \tau_{21} + \epsilon_{g2} E_{bg}).$$

The appropriate data for substitution in these equations are

$$F_{21} = 1,$$

$$F_{12} = F_{21}A_2/A_1 = 0\cdot429,$$

$$F_{11} = 1 - F_{12} = 0\cdot571,$$

$$\tau_{11} = e^{-1\cdot16} = 0\cdot3135,$$

$$\tau_{12} = e^{-0\cdot23} = 0\cdot8 = \tau_{21},$$

$$\epsilon_{g1} = 1 - e^{-0\cdot24} = 0\cdot213,$$

$$\epsilon_{g2} = 1 - e^{-0\cdot236} = 0\cdot210,$$

$$E_{b1} = 56\cdot7 \times 1\cdot1^4 = 83\cdot01 \text{ kW/m}^2,$$

$$E_{b2} = 56\cdot7 \text{ kW/m}^2,$$

$$E_{bg} = 161\cdot9 \text{ kW/m}^2,$$

$$0\cdot928W_1 - 0\cdot1373W_2 = 63\cdot62,$$

$$-0\cdot24\,W_1 + W_2 = 49\cdot9$$

and
$$W_2 = 68\cdot8 \text{ kW/m}^2;$$

$$Q = 12 \times 0\cdot7 \times (68\cdot8 - 56\cdot7)/0\cdot3$$

$$= 339 \text{ kW}.$$

4.3. EXCHANGE AREAS FOR ABSORBING MEDIA

The concept of exchange areas, introduced in Section 2.2, can be extended to enclosures containing absorbing media. Equation (2.5) gives the amount of radiation emitted from a small area of a grey surface (ΔA) at an angle ϕ and transmitted through a solid angle Δw. If the radiation is transmitted through an absorbing medium, it will be attenuated by $\tau = e^{-Kr}$ where r is the distance travelled by the radiation and the expression becomes

$$\Delta Q = \epsilon E_b \Delta A \cos \phi \, \Delta w e^{-Kr}/\pi. \qquad (4.38)$$

As in Section 2.2, this equation may be expressed in differential form and integrated to give the total transmission of energy from one surface (A_1) to another (A_2) (Fig. 4.7a):

$$Q = \epsilon_1 E_{b1} \int^{A_1} \int^{A_2} \cos \phi_1 \cos \phi_2 \, e^{-Kr} dA_1 dA_2 / \pi r^2 \qquad (4.39)$$

$$= \epsilon_1 E_{b1} \overline{s_1 s_2} \qquad (4.40)$$

where $\overline{s_1 s_2} = \int^{A_1} \int^{A_2} \cos \phi_1 \cos \phi_2 \, e^{-Kr} dA_1 dA_2 / \pi r^2.$ $\qquad (4.41)$

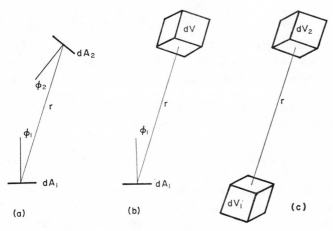

Fɪɢ. 4.7.

Comparison with eqs. (2.9) and (2.10) shows that $\overline{s_1 s_2}$ is equivalent to the exchange area $\overline{12}$. From the symmetry of eq. (4.41):

$$\overline{s_1 s_2} = \overline{s_2 s_1}. \qquad (4.42)$$

The total radiation emitted by A_1 is $A_1 \epsilon_1 E_{b1}$ and the fraction incident upon surface 2 is

$$Q/A_1 \epsilon_1 E_{b1} = \overline{s_1 s_2}/A_1 \qquad (4.43)$$

and

$$\overline{s_1 s_2} = F_{12} \tau_{12}$$

$$= A_1 F_{12} e^{-KL_{12}}. \qquad (4.44)$$

Similarly, it may be shown that the radiation which is emitted by a surface and absorbed by a volume of gas (V) is (Fig. 4.7b)

$$Q = \epsilon_1 E_{b1} \int^{A_1} \int^{V} \cos \phi_1 \, Ke^{-Kr} dV dA_1 / \pi r^2 \qquad (4.45)$$

and a surface–gas exchange area $(\overline{s_1 g})$ may be defined as

$$\overline{s_1 g} = \int^{A_1} \int^{V} \cos \phi_1 \, Ke^{-Kr} dV dA_1 / \pi r^2. \qquad (4.46)$$

The fraction of radiation absorbed by the gas is

$$a_{1g} = \epsilon_{1g} = \overline{s_1 g} / A_1. \qquad (4.47)$$

Hottel and Sarofim (1967) have evaluated $\overline{s_1 s_2}$ and $\overline{s_1 g}$ for some geometries and the radiosity calculations for an enclosure can be conveniently carried out using exchange areas instead of mean beam lengths.

It is also possible to define a gas–gas exchange area on the basis of the differential volumes shown in Fig. 4.7c:

$$\overline{g_1 g_2} = \int^{V_1} \int^{V_2} K_1 K_2 e^{-Kr} dV_1 dV_2 / \pi r^2 \qquad (4.48)$$

where $\overline{g_1 g_2}$ describes the exchange of radiation between two volumes of gas separated by a distance r. However, if the conceptual surfaces suggested in Section 3.6 are used to separate isothermal gas zones, the transfer of radiation from one gas zone to another is considered to take place in two steps, that is, transfer from the gas zone to the separating surface and then from that surface to the next gas zone. Consequently, it is not necessary to invoke gas–gas exchange areas if this procedure is adopted.

CHAPTER 5

Radiative Heat Transfer Applications

5.1. FURNACES

5.1.1. *Introduction*

A furnace is an enclosure in which stock is heated to a high temperature; examples are glass and steel melting furnaces, cement kilns and annealing furnaces. The heat required for the process may be produced either by combustion of a fuel or electrically by resistive, inductive, capacitative or arc heating.

Furnaces are commonly classified into two categories: batch and continuous. In a batch furnace, the stock is placed in the enclosure, heated to the required temperature and then removed. In a continuous furnace, the stock is moved through the enclosure as it is heated. There are, however, a large number of different types of furnace, varying in geometry and heating arrangement within each classification (Thring, 1962; Trinks and Mawhinney, 1961; and *Efficient Use of Fuel*, 1958).

In high-temperature furnaces, that is, above about 800°C (1073 K), radiation can be the dominant mode of heat transfer and a considerable effort has been expended in developing methods of estimating rates of radiative heat exchange in furnaces. The methods are also valid for lower temperatures but, in those cases, the combined effects of convective and radiative heat transfer must be taken into account. The calculations enable furnace performance to be predicted rapidly and reliably without using experimental furnaces, thus avoiding the inherently large costs of the latter. Calculations of this type are frequently termed mathematical models, as opposed to an experimental or physical model.

81

Although the principles of radiative heat transfer are well understood, a rigorous furnace calculation is not possible at present for two reasons. Firstly, there are insufficient experimental data on emissivities and absorptivities and, secondly, the computational requirements are too large. Consequently, simpler models are used and the validity of the assumptions made are tested by comparison with furnaces already in use.

The techniques for determining heat transfer in an enclosure containing an absorbing medium have been described in Chapter 4. In this chapter, the methods of treating a real furnace are described. The differences between a real furnace and the enclosures so far considered are:

1. The furnace geometry may be complex (Section 5.1.2).
2. The temperature of the stock at a particular point in the furnace may vary with time, as in batch furnaces (Section 5.1.3).
3. The temperature may vary considerably throughout the furnace, as in continuous furnaces (Section 5.1.4).
4. The gases are not grey, an assumption inherent in the concept of mean beam length (Section 5.1.5).
5. The surfaces are not grey (Section 5.1.6).

5.1.2. *Furnace Geometry*

The enclosures considered in Chapters 3 and 4 comprised plane or cylindrical surfaces. A number of other real geometries, for example, the curved roof of some furnaces, may be treated simply as if they were plane surfaces.

In Fig. 5.1, A_1 is the area of the concave surface and A_2 is the area of the plane surface which is to replace A_1. The radiant energy from the curved surface transmitted across A_2 is

$$A_2 H_2 = A_1 F_{12} W_1$$
$$= A_2 F_{21} W_1$$
$$= A_2 W_1. \qquad (5.1)$$

The radiosity equation for A_1, based on emission of radiation from A_1 only, is

$$W_1 = \epsilon_1 E_{b1} + \rho_1 F_{11} W_1$$
$$= \epsilon_1 E_{b1}/(1 - \rho_1 F_{11}). \tag{5.2}$$

Substitution in eq. (5.1) gives

$$A_2 H_2 = A_2 \epsilon_1 E_{b1}/(1 - \rho_1 F_{11}).$$

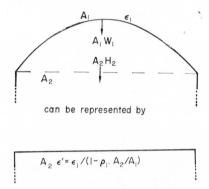

can be represented by

$$A_2 \; \epsilon' = \epsilon_1 /(1 - \rho_1 \cdot A_2/A_1)$$

FIG. 5.1. Radiation incident upon and leaving a concave surface.

As far as emission of radiation is concerned, the concave surface may be replaced by a plane surface of area A_2, a black-body emissive power of E_{b1} and an effective emissivity of

$$\epsilon' = \epsilon_1/(1 - \rho_1 F_{11}) \tag{5.3}$$

where $F_{11} = 1 - A_2/A_1$.

Both batch and continuous furnaces may be heated by radiant elements, which may be either electrical resistors or gas-fired tubes (Fig. 5.2). The analysis of heat transfer from an array of elements of this type is laborious if each element is treated as a separate surface. However, if the length of the furnace and the length of the elements are much greater than their diameter and separation, the array can be replaced, for the purposes of calculation, by a grey plane.

E.C.—D

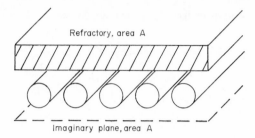

FIG. 5.2. An array of elements backed by a refractory.

EXAMPLE 5.1

Calculate the effective emissivity of an array of resistors backed by refractory. They are of circular cross-section, 45 mm in diameter, 1·5 m long and spaced 110 mm apart (centre to centre). The refractory may be assumed to be totally reflecting and the resistors to be grey with an emissivity of 0·75.

In this solution, it is assumed that the radiosity is constant over the surface of the resistors and that the array extends to infinity. The effective emissivity ϵ' of this arrangement is obtained by dividing the total radiation (Q) through an imaginary plane, of area (A) equal to that of the refractory, by the black-body emission from the same plane at the temperature of the resistors (Fig. 5.2).

Thus,
$$\epsilon' = Q/A\sigma T^4. \tag{a}$$

Since the resistors are much longer than their diameter and separation, they may be treated as infinitely long and the calculations carried out for unit length.

Nomenclature:

H_1 incident flux on the refractory,
W_1 total leaving flux (radiosity) from the refractory,
H_2 incident flux on a single resistor,
W_2 leaving flux from a resistor,
F_{22} view factor from one resistor to the next,
F_{21} view factor from one resistor to refractory (or plane)
        $(= F_{23})$,

F_{13} view factor from refractory to plane,
N total number of resistors,
C resistor spacing (A/N),
A_2 area of resistor per unit length.

The total energy passing through the imaginary plane is

$$Q = AF_{13}W_1 + NA_2F_{23}W_2$$
$$= AF_{13}W_1 + AA_2F_{21}W_2/C \qquad \text{(b)}$$

and the problem becomes one of calculating W_1 and W_2.
The radiosity equations are

1. For the refractory:

$$AW_1 = NA_2F_{21}W_2. \qquad \text{(c)}$$

2. For a single resistor:

$$A_2W_2 = A_2\epsilon_2E_{b2} + \rho_2(2A_2F_{22}W_2 + AF_{12}W_1). \qquad \text{(d)}$$

Equations (c) and (d) may be solved for W_2 and W_1 and substitution into eq. (b) gives

$$Q = \frac{A\sigma T^4 \epsilon_2 A_2 F_{21}(F_{13} + 1)}{C(1 - \rho_2[A_2F_{21}^2/C + 2F_{22}])}.$$

From eq. (a), the effective emissivity of the array is

$$\epsilon' = \frac{\epsilon_2 A_2 F_{21}(F_{13} + 1)}{C(1 - \rho_2[A_2F_{21}^2/C + 2F_{22}])}. \qquad \text{(e)}$$

The view factor, F_{22}, may be evaluated by the crossed-string method (Section 2.2.1.3):

$$F_{22} = 0 \cdot 066.$$

Since each resistor can "see" two neighbouring resistors and two planes,

$$2F_{22} + 2F_{21} = 1$$

and

$$F_{21} = 0 \cdot 434.$$

The refractory can "see" N resistors and the imaginary black plane so that

$$F_{13} + NF_{12} = 1,$$

$$F_{13} = 1 - NA_2F_{21}/A$$

$$= 1 - A_2F_{21}/C$$

$$= 1 - 45\pi \times 0{\cdot}434/110$$

$$= 0{\cdot}442.$$

From eq. (e)

$$\epsilon' = 0{\cdot}887\epsilon_2$$

$$= 0{\cdot}666.$$

Arrays of gas-fired radiant tubes or boiler tubes may be treated in a similar manner. In practice, the temperatures of the elements or tubes may vary along the furnace. This temperature variation may be represented by means of the zoning approach discussed in Section 3.6.

5.1.3. *Variation of Temperature with Time*

In a batch furnace, the stock temperature changes with time. The simplest analysis of heat transfer in this type of furnace is the well-stirred model in which the combustion products filling the enclosure are treated as being isothermal and of constant composition and the wall and roof temperatures are considered constant (Example 4.5). In a fuel-fired furnace with stock on the floor, the walls and roof may be treated as a single totally reflecting surface. Alternatively, if there is substantial heat loss through the walls or roof and if their emissivities and temperatures are different, they may be treated as separate isothermal surfaces. If the heating is by exposed electrical resistors or gas-heated radiant tubes, they can be treated as a plane surface as shown in Example 5.1. The calculation of heat transfer to the stock is then essentially the same as in Example 4.5. The rate of change of stock temperature may be calculated by expressing the heat transfer to the stock as a function of stock temperature and equating this to the rate of change of enthalpy with time. Convective heat transfer contributions can be included if necessary.

EXAMPLE 5.2

The combustion products in a fuel fired furnace, 6 m long and 2 m wide, can be assumed to be isothermal at a temperature of 1300 K and emit and absorb as a grey gas with an absorption coefficient of 0.2 m^{-1}. The walls and roof are at a temperature of 1100 K and have an emissivity of 0.6. Steel stock on the floor of the furnace has an effective emissivity of 0.7, a thermal capacity of 3900 kJ/m^3 K and an effective thickness of 0.25 m. The distance from the surface of the stock to the roof of the furnace is 1 m. Given that the initial temperature of the stock is 300 K, calculate the time required to heat the stock to 1000 K.

From Example 4.5, the radiosity equations are

$$0.928W_1 - 0.137W_2 = 63.62$$

$$-0.24W_1 + W_2 = 0.7E_{b2} + 0.3 \times 0.21 \times 161.9.$$

Solving for W_2:

$$W_2 = 27.7 + 0.726E_{b2} \text{ kW/m}^2$$

and the net heat flux to the stock is

$$q = 0.7 \times (27.7 + 0.726E_{b2} - E_{b2})/0.3$$

$$= 64.6 - 0.639E_{b2}$$

$$= 64.6 - 36.25(T/1000)^4 \text{ kW/m}^2.$$

The rate of change of enthalpy of the stock is

$$dH/dt = d(3900 \times 02.5T)/dt$$

$$= 975dT/dt$$

$$= q$$

or

$$975dT/dt = 64.6 - 36.25(T/1000)^4$$

and the time required for the stock to reach 1000 K is given by

$$t = \int_{300}^{1000} \frac{975\,dT}{64{\cdot}6 - 36{\cdot}25(T/1000)^4} \text{ sec}$$

$$= \int_{300}^{1000} \frac{0{\cdot}271\,dT}{64{\cdot}6 - 36{\cdot}25(T/1000)^4} \text{ hr.}$$

This equation is solved graphically by measuring the area under the curve in Fig. 5.3 to give

$$t = 3{\cdot}65 \text{ hr.}$$

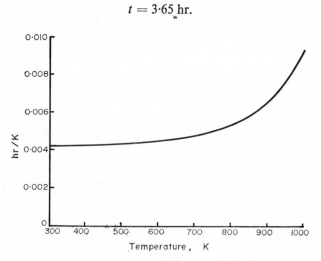

FIG. 5.3. Graphical solution for Example 5.2.

5.1.4. *Variation of Temperature within the Furnace*

In a continuous furnace, stock enters one end at a low temperature and emerges at a high temperature. In some applications, it may be cooled after processing within the furnace. Thus, there are substantial changes in surface and gas temperatures along the furnace, and the simplest model to use in this case is the plug flow model. The gases are assumed to flow along the furnace as a plug with constant transverse temperatures and compositions. The furnace is divided into sections as in Fig. 5.4 and each surface and the gas volume within each section is assumed to be isothermal. The length of the individual sections depends upon the closeness with which the temperature gradient in the furnace is to be modelled. The simplest form of calculation considers each section to be thermally isolated, that is, the boundaries between each section and its neighbours are considered to be adiabatic walls across which there is no net heat transfer. This model is usually called the long chamber or long furnace model. If the side walls are totally reflecting refractories, the four walls of the section may be treated as one surface. It may also be possible to treat the roof in the same way. The analysis is now similar to the well-stirred model, a separate calculation being performed for each section.

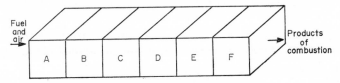

FIG. 5.4. Long furnace model (Example 5.3).

EXAMPLE 5.3

Calculate the equilibrium temperature distribution in a pulverized fuel furnace in which heat transfer to the walls is considered to take place mainly by radiation. The furnace is 6 m long and of 1 × 1 m square cross-section. The mass of fuel and air entering the furnace is 2 kg/s and its temperature is 400 K. Assume that the behaviour of the furnace can be represented by six zones (Fig. 5.4) and that the two end faces of the furnace and the side walls are black and cold except for the four side walls of the first zone which are completely reflecting refractory. Combustion of the fuel provides 4000 kW distributed throughout the furnace as shown in the following table which also lists the absorption coefficients of the products in the six zones.

	A	B	C	D	E	F
Heat release by combustion, kW	1160	2140	420	160	120	0
Absorption coefficient, m^{-1}	2·00	1·75	0·80	0·50	0·40	0·30

The following form of solution has been used by Field *et al.* (1967) and compared favourably with a solution based on radiative transfer between zones. The latter solution utilized the method due to Hottel and Cohen (1958).

The assumptions made in this solution are

1. No variation of temperature and properties occur over the cross-section.
2. Radiation from the combustion products in each zone occurs at a level corresponding to $\epsilon\sigma(T_i^4 + T_o^4)/2$ where T_i and T_o are the temperatures of the inlet and outlet faces of the zone.
3. The reactants and products passing through the furnace have a specific heat of 1·26 kJ/kg.

The mean beam length of a cube of absorbing medium is $0.6L$ where L is the length of the side (Table 4.1) and the emissivities of the products in the six zones are

	A	B	C	D	E	F
Emissivity	0·70	0·65	0·38	0·26	0·215	0·165

A heat balance on each zone gives

heat generated by combustion = radiative heat transfer to walls + rise in enthalpy of the products passing through the furnace,

$$Q_C = Q_R + Q_S$$
$$= A\epsilon\sigma(T_i^4 + T_o^4)/2 + mc_p(T_o - T_i)$$

where m is the mass flow rate (kg/s) and c_p the mean specific heat at constant pressure of the products.

For the first zone:

$$1160 = 1 \times 0.70 \times 56.7 \times 10^{-12} \times (T_i^4 + T_o^4)/2$$
$$+ 2 \times 1.26 \times (T_o - T_i),$$
$$T_o + 7.88 \times 10^{-12}T_o^4 = 460 + T_i - 7.88 \times 10^{-12}T_i^4.$$

This equation may be solved graphically to give $T_o = 850$ K. Repeating the procedure for each zone in turn provides the following values for the temperatures in the furnace. (Note: the area (A) in zones B to E is 4 m².)

	A	B	C	D	E	F
Inlet temperature K	400	850	1525	1500	1450	1420
Outlet temperature K		850	1525	1500	1450	1420	1410

The analysis may be improved by removing the concept of the hypothetical adiabatic walls and using the zoning approach discussed in Section 3.6, with the addition of transmissivity and gas emissivity terms.

For both batch and continuous furnaces, the analysis can be further improved by dividing the surfaces into smaller isothermal zones.

5.1.5. *Representation of Real Gases*

In order to use the concept of a mean beam length, it is necessary to assume that the gases or flames are grey. Figure 5.5 shows the variation of emissivity with pL for a grey gas and also the variation of the

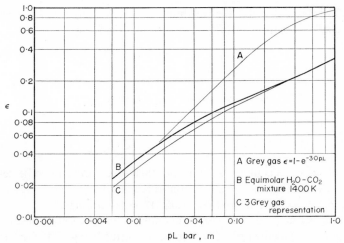

FIG. 5.5. Variation of emissivity with pL.

emissivity of an atmosphere containing equi-molal concentrations of water vapour and carbon dioxide at a temperature of 1400 K obtained from data presented by Hadvig (1970). Hadvig has noted that the combustion products of petroleum fuels contain carbon dioxide and water vapour in approximately equi-molal quantities and that those of methane contain twice as much water. Consequently, he has calculated the emissivities of atmospheres containing carbon dioxide and water vapour in these proportions from the data of Hottel (Figs. 4.2 to 4.6) and these values are presented in Appendix 5. The assumption of greyness is adequate for a simple calculation but, for greater accuracy, the departure from greyness must be considered. This can be done by equating the emissivity of the real gas with a weighted sum of the emissivities of a number of grey gases:

$$\epsilon_g = a_1(1 - e^{-k_1 pL}) + a_2(1 - e^{-k_2 pL}) + a_3(1 - e^{-k_3 pL}) + \ldots$$

$$= \sum a_i - \sum a_i e^{-k_i pL} \tag{5.4}$$

where $k_1 < k_2 < k_3 < \ldots$.

At large values of pL, e^{-kpL} is very small and

$$\sum a_i = \epsilon_g \text{ at large } pL. \tag{5.5}$$

In order to determine the weighting coefficients (a_i) and the attenuation coefficients (k_i), eq. (5.4) is rewritten in the form

$$\ln(\sum a_i - \epsilon_g) = \ln(\sum a_i e^{-k_i pL}). \tag{5.6}$$

Since $k_1 < k_2 < k_3 < \ldots$, then at large pL,

$$e^{-k_1 pL} \gg e^{-k_2 pL} \gg e^{-k_3 pL} \gg \ldots$$

and eq. (5.6) becomes

$$\ln(\sum a_i - \epsilon_g) = -k_1 pL + \ln(a_1). \tag{5.7}$$

If $\ln(\Sigma a_i - \epsilon_g)$ is plotted against pL, then $\ln(a_1)$ and k_1 can be determined by drawing the tangent to the curve at large pL, and measuring its gradient (k_1) and its intercept ($\ln(a_1)$) on the ordinate.

a_2 and k_2 may be obtained in a similar way by plotting $\ln(\Sigma a_1 - \epsilon_g - a_1 e^{-k_1 pL})$ against pL. The number of terms in eq. (5.4) for which coefficients are calculated depends on the accuracy of fit required.

EXAMPLE 5.4

Represent the emissivity of a furnace atmosphere containing an equimolal mixture of carbon dioxide and water vapour ($p = p_{CO_2} + p_{H_2O} = 0.30$ bar) at 1400 K by means of the weighted sum of the emissivities of three grey gases.

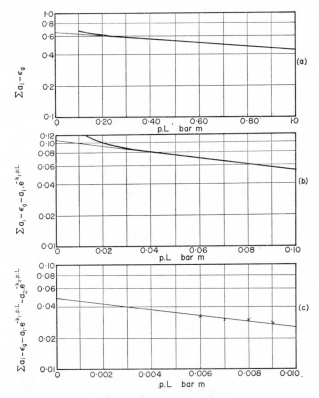

FIG. 5.6. Evaluation of weighting and attenuation coefficients (Example 5.4).

The variation of the emissivity of the atmosphere with pL is obtained from Appendix 5 and is plotted in Fig. 5.5. By extrapolating the curve to high values of pL, a value of $\Sigma a_i = 0.8$ is obtained. $\text{Ln}(\Sigma a_i - \epsilon_g)$ in the range 0.1 to 1.0 bar m is plotted in Fig. 5.6a. The gradient of the tangent to the curve and the intercept give

$$a_1 = 0\cdot64$$

and
$$k_1 = \ln(0\cdot64/0\cdot46)$$
$$= 0\cdot327 \text{ bar}^{-1}\text{m}^{-1}.$$

Since $a_1(1 - e^{-k_1 pL})$ now approximates to the emissivity of the atmosphere in the range $0\cdot1$ to $1\cdot0$ bar m, it is necessary to plot $\log(\Sigma a_i - \epsilon_g - a_i e^{-k_1 pL})$ in the range $0\cdot01$ to $0\cdot10$ bar m in order to evaluate a_2 and k_2. This has been done in Fig. 5.6b and gives

$$a_2 = 0\cdot105$$

and
$$k_2 = 10 \ln(0\cdot105/0\cdot064)$$
$$= 4\cdot95 \text{ bar}^{-1}\text{m}^{-1}.$$

Similarly, a_3 and k_3 are evaluated by plotting

$$\ln(\textstyle\sum a_i - \epsilon_g - a_1 e^{-k_1 pL} - a_2 e^{-k_2 pL})$$

in the range $0\cdot001$ to $0\cdot01$ (Fig. 5.6c) and gives

$$a_3 = 0\cdot046$$

and
$$k_3 = 100 \ln(0\cdot046/0\cdot025)$$
$$= 60 \text{ bar}^{-1}\text{m}^{-1}.$$

A number of values of the weighted sum of the three grey gases for which coefficients have been calculated have been plotted in Fig. 5.5 to show the kind of fit that is obtained over the range of pL used.

The number of terms necessary to produce a good fit depends on the range of pL over which representation is required and upon the shape of the ϵ versus pL curve for the real gas.

The procedure can be repeated at different temperatures to allow the variation of the coefficients with temperature to be determined. In practice, the values of k are assumed to be independent of temperature, and temperature dependence is described solely by the weighting coefficients (a_i).

The absorptivity of a real gas may also be represented by the absorptivities of a number of grey gases, but, in this case, the coefficients are functions of the temperature of the source of radiation as well as of gas temperature (Sarofim, 1962).

A similar procedure has been used for fitting a series of grey gases to experimental data for luminous flames (Beer and Johnson, 1972).

As far as calculation of heat transfer within enclosures is concerned it is necessary to express the emissivities and transmissivities in the radiosity equations as

$$\epsilon_g = a_1(1 - e^{-k_1 p L}) + a_2(1 - e^{-k_2 p L'}) + \ldots,$$

$$\tau = 1 - a_g = a_1' e^{-k_1 p L'} + a_2' e^{-k_2 p L'} + \ldots.$$

a_i' is different from a_i because the temperature of radiation passing through the gas must be taken into account when representing the absorptivity and L' is different from L for the reasons discussed in Section 4.2.

5.1.6. Heat Transfer between Real Surfaces

If the surfaces in an enclosure cannot be considered to be grey, the calculation of heat transfer becomes more complex because spectral emissivities must be incorporated in the calculations (Kreith, 1962; Wiebelt, 1966; Hottel and Sarofim, 1967; Touluokian, 1972). The radiosity equations for an enclosure comprising non-grey surfaces are of the form

$$\int_0^\infty W_{i\lambda} \, d\lambda = \int_0^\infty (\epsilon_{i\lambda} E_{b i\lambda} + \sum \rho_{j\lambda} W_{j\lambda} F_{ij}) d\lambda \qquad (5.8)$$

and the net heat transfer to surface i is

$$Q_i = A_i \int_0^\infty \epsilon_{i\lambda}(W_{i\lambda} - E_{i\lambda})/\rho_{i\lambda}. \qquad (5.9)$$

The set of simultaneous equations [eq. (5.8)] can only be solved with great difficulty especially if non-Lambert behaviour is also considered.

The Monte Carlo method, outlined in Section 2.2.1.4, offers an alternative method of solution. As in Example 2.7, the coordinates of a point of emission of a beam of radiation are chosen on one surface. The angles of emission are chosen in such a way as to account for any deviation from Lambert's cosine law and the wavelength of emission selected to account for deviations from greyness, according to the procedure outlined in Appendix 2. Random numbers are also used to determine if the beam is absorbed or reflected when it strikes a second surface. If reflected, the path of the beam is followed until it is eventually absorbed. The total number of beams emitted from any one surface is proportional to the emissive power of that surface (ϵE_b) and each surface is considered in turn. The net heat transfer to any one surface is the number of beams absorbed minus the number of beams emitted.

The advantages of the Monte Carlo method are its simplicity of formulation for the purpose of machine computation and the fact that it is relatively easy to take the presence of a spectrally absorbing medium into account. Steward and Cannon (1970) and Taniguchi (1969) have described computations based on this method for the determination of heat transfer in furnaces.

5.2. SOLAR RADIATION

Calculations of the amount of solar energy incident upon surfaces on the earth are carried out either to estimate

1. the contribution made by solar energy to the heating of buildings which are otherwise heated by other forms of energy (Jones, 1967; Bassett and Pritchard, 1970) or
2. the amount of solar energy which can be collected to provide a significant proportion of, if not all, the heating requirements of a consumer (Brinkworth, 1972).

5.2.1. *Solar Radiation upon Surfaces on Earth*

In both cases referred to above, calculations of the quantities of solar energy incident upon horizontal and non-horizontal surfaces are required.

1. *Horizontal surfaces*

In Section 1.7, the solar energy incident upon a horizontal surface was calculated for two cases,

(a) a value at a selected time of day for which attenuation by the earth's atmosphere was taken into account and
(b) the total quantity of energy received during the day.

The latter calculation, in which attenuation by the Earth's atmosphere was neglected, was based on eq. (1.28) which involves the assumption that the Earth is at a constant distance from the Sun. In fact, the path of the Earth is an ellipse with the Sun at one of the foci. The eccentricity (e) of the ellipse is very small and the orbit of the Earth is almost circular with its semi-major axis (a) equal to 149.5×10^6 km. The longest and shortest Earth–Sun distances which occur on 1 July and 1 January are called the aphelion and perihelion distances (R_a and R_p) respectively.

Since $e = 0.01673$,

$$R_p = a(1 - e) = 147.1 \times 10^6 \text{ km},$$
$$R_a = a(1 + e) = 152.1 \times 10^6 \text{ km}.$$

The quantity of solar energy arriving at the Earth is inversely proportional to the square of the Earth–Sun distance, a relationship which has been used to produce the ratios of the actual quantity of solar radiation arriving (G_t) to the solar constant (G_0) given in Fig. 5.7.

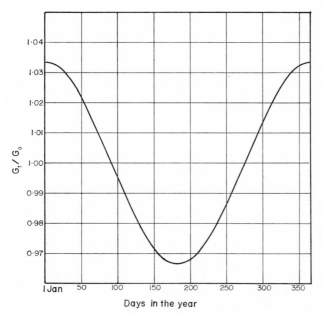

FIG. 5.7. Intensities of solar radiation relative to the solar constant.

Equation (1.28) can be rewritten to give the amount of solar energy incident upon a horizontal surface on the earth, neglecting the effects of the Earth's atmosphere:

$$G = 2 \cdot 75 \times 10^4 (G_t/G_0) G_0 (H \sin (l) \sin (d) + \cos (l) \cos (d) \sin (H)).$$
$$(5.10)$$

At sunrise, $z = \pi/2$ and $\cos (z) = 0$ so that, from eq. (1.24),

$$\sin (l) \sin (d) = -\cos (l) \cos (d) \cos (H). \qquad (5.11)$$

Substituting this expression into eq. (5.7):

$$G = 2 \cdot 75 \times 10^4 (G_t/G_0) G_0 \cos (l) \cos (d) (\sin (H) - H \cos (H)).$$
$$(5.12)$$

The results of calculations based on this equation are presented graphically for the whole year and for all latitudes in Fig. 5.8.

2. *Vertical surfaces*

In Section 1.17 it was shown that the solar radiation incident upon a horizontal surface is $G_0 \cos (z)$. This is the normal component of the Sun's radiation, the horizontal component is $G_0 \sin (z)$. Similarly, the radiation incident upon a vertical surface directly facing the sun is

$$q = G_0 \sin (z). \qquad (5.13)$$

If the vertical surface does not face the Sun directly but is inclined at an angle n to the rays from the Sun as shown in Fig. 5.9, the amount of solar radiation incident upon the surface is

$$q = G_0 \sin (z) \cos (n). \qquad (5.14)$$

The angle n is the difference between (1) the azimuthal angle (az) which defines the Sun's position relative to due south and (2) the direction of the normal to the surface relative to due south (Fig. 5.9).

It can be shown that

$$\sin (az) = \cos (d) \sin (h)/\sin (z). \qquad (5.15)$$

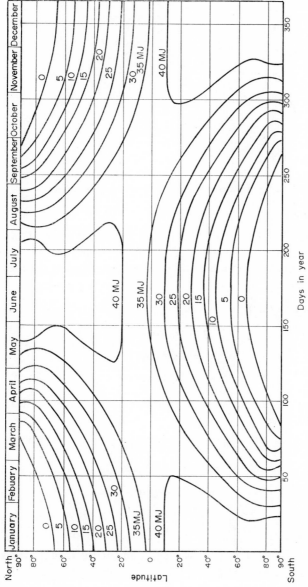

FIG. 5.8. Daily amounts of solar energy arriving at the earth (neglecting effects of the Earth's atmosphere).

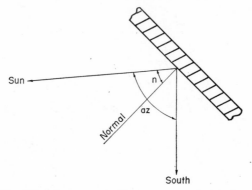

Fig. 5.9. Solar energy incident upon a vertical surface.

EXAMPLE 5.5

Calculate the amount of solar radiation incident upon (1) a horizontal surface and (2) a vertical surface facing south-west for a zenith angle of 25° and an azimuth of 80° west of south. Neglect the effects of the atmosphere.

1. From eq. (1.22):

$$q = G_0 \cos (z)$$
$$= 1 \cdot 39 \cos (25)$$
$$= 1 \cdot 26 \text{ kW/m}^2.$$

2. Figure 5.9 illustrates solar radiation arriving at a vertical wall facing south-west.

$$q = G_0 \sin (z) \cos (n)$$
$$= G_0 \sin (25) \cos (35)$$
$$= 0 \cdot 481 \text{ kW/m}^2.$$

3. *Tilted surfaces*

If a surface is tilted from the vertical at an angle β, then the radiation incident upon it will comprise

1. a component of the radiation on the corresponding vertical surface, G_0 sin (z) cos (n) cos (β) and
2. a component of the radiation on the corresponding horizontal surface, G_0 cos (z) sin (β).

The total amount of radiation is

$$q = G_0 \left(\cos (z) \sin (\beta) \pm \sin (z) \cos (n) \cos (\beta)\right). \qquad (5.16)$$

When the surface faces the Sun, the second term in eq. (5.13) is added, when tilted away it is subtracted (Fig. 5.10).

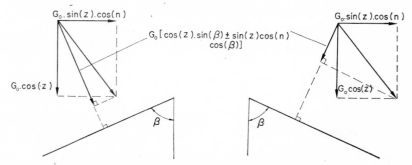

FIG. 5.10. Solar energy incident upon a surface inclined to the vertical.

Equation (5.13) is the same as eq. (1.22) with i, the angle the Sun's rays makes with the normal to the surface, defined by

$$\cos (i) = \cos (z) \sin (\beta) \pm \sin (z) \cos (n) \cos (\beta). \qquad (5.17)$$

EXAMPLE 5.6

Calculate the amount of solar radiation incident upon a wall facing south-west and tilted 60° to the vertical. The zenith angle and azimuth are 25° and 80° west of south respectively. Determine the amount of radiation for such a surface tilted towards the Sun and away from it.

1. Tilted towards the Sun:

$$q = G_0 \left(\cos (25) \sin (60) + \sin (25) \cos (35) \cos (60)\right)$$

$$= G_0(0 \cdot 785 + 0 \cdot 174)$$

$$= 1 \cdot 43 \text{ kW/m}^2.$$

2. Tilted away from the Sun:

$$q = 0{\cdot}611G_0 = 0{\cdot}848 \text{ kW/m}^2.$$

5.2.2. Absorption of Solar Radiation by Surfaces

As far as utilization of solar energy is concerned, the important factor is not only how much solar energy is incident upon a surface but also how much can be absorbed. The extent to which solar energy is absorbed depends on the properties of the surface which will also determine the equilibrium temperature of the surface. The latter temperature is also affected by changes in the system of collection of solar energy as is illustrated by the following calculations.

Flat-plate collector

Consider a flat-plate lying on an insulated base. The only significant heat transfer is by solar radiation to the plate and by emission from the plate (neglecting radiative transfer from the surroundings and convective effects).

$$\alpha_s G = \epsilon \sigma T^4 \tag{5.18}$$

where α_s is the absorptivity of the surface to solar radiation, G the intensity of solar radiation and T the temperature of the surface.

$$T = (\alpha_s G / \sigma \epsilon)^{\frac{1}{4}}. \tag{5.19}$$

High values of α_s/ϵ and α_s are required to obtain the highest possible temperature and to absorb the largest quantity of radiation respectively. Although some materials—metals, for example—have high values of α_s/ϵ, their absorptivities are low. Surfaces of high absorptivity are also characterized by ratios of α_s/ϵ of about 1 and these are called neutral absorbers.

The simple flat-plate collector can be improved by a technique which effectively increases α_s/ϵ without reducing α_s too much. This is achieved by coating a polished metal surface with a thin deposit of a black oxide, that of nickel or copper. This deposit has a high absorptivity for the short-wave solar radiation but, for layers thinner than the longer wavelengths corresponding to the absorber temperatures, the absorptivity (and the emissivity) is close to that of the metal below.

Greater equilibrium temperatures can be achieved by reflecting more solar radiation on to the absorbing surface by means of plane mirrors or a parabolic mirror (Kreith, 1962; Brinkworth, 1972). The factor by which radiation is increased is called the concentration ratio. Temperatures of the order of 2000 K may be obtained by this means and the solar furnace is based on this kind of arrangement.

Example 5.7

Compare the equilibrium temperature of a flat plate, insulated from the Earth and subjected to solar radiation of 0.8 kW/m^2, for the following conditions.

1. The plate is black.
2. The plate is of polished metal coated with a thin layer of cuprous oxide, the emissivity of which can be considered to be 0.92 at wavelengths below $2.0 \ \mu\text{m}$ and 0.1 at greater wavelengths.
3. A parabolic mirror is used to concentrate the solar radiation by a ratio of 1000.

$$(1) \quad T = (0.8/56.7)^{\frac{1}{4}} \times 1000$$

$$= 345 \text{ K} \quad \text{or} \quad 72°\text{C}.$$

$$(2) \quad T = (0.92 \times 0.8/56.7 \times 0.1)^{\frac{1}{4}} \times 1000$$

$$= 600 \text{ K} \quad \text{or} \quad 327°\text{C}.$$

$$(3) \quad T = (800/56.7)^{\frac{1}{4}} \times 1000$$

$$= 1938 \text{ K} \quad \text{or} \quad 1665°\text{C}.$$

Another improvement to the design of a flat-plate solar collector is to fit one or more transparent cover plates above the absorbing surface in order to take advantage of the "greenhouse" effect (Example 1.7). In Example 5.8, the properties of a glass cover are represented more precisely than in Example 1.7 and the temperatures of both the glass and the absorbing surface estimated.

EXAMPLE 5.8

A thin plate is insulated from the Earth and shielded from solar radiation by a glass sheet. Calculate the temperatures of the glass and plate when the solar radiation incident upon them is 0.8 kW/m². The plate is grey with an absorptivity of 0.95 and the properties of the glass may be represented by the following data:

Wavelength	0.4 μm	0.4 to 2.5 μm	2.5 μm
Absorptivity	0·2	0	0·9
Reflectivity	0·8	0·1	0·1
Transmissivity	0	0·9	0

This problem can be solved by

1. determining how much solar energy is incident on the glass within the three spectral regions defined by 0.4 and 2.5 μm;
2. setting up balances for the fluxes leaving the glass and plate for each of the spectral regions; and
3. setting up heat balances for the glass and plate over the whole thermal range.

1. The percentages of the radiation emitted by a black body within given wavelength ranges can be obtained from the curve in Fig. 1.5 but the values are not sufficiently accurate for the high values of λT encountered in this example. Consequently, a further set of values of $D_{0\lambda}$ are tabulated in Appendix 4. The amounts of radiation emitted in the three regions of the spectrum considered in this example are:

Wavelength	<0.4 μm	0.4 to 2.5 μm	>2.5 μm
D_0 %	12·5	84·1	3·4
$D_0 \times G$ kW/m²	0·10	0·6728	0·0272

2. *Leaving fluxes.* The radiation incident upon and leaving the glass and the plate surfaces within the three spectral regions is shown in Fig. 5.11. The temperatures of the glass and plate are unlikely to be greater than about 150°C, and less than 0·1 per cent of the total radiation emitted by bodies at temperatures lower than this occurs at wavelengths below 2.5 μm. Consequently, the flux leaving the glass and plate is negligible at wavelengths less than 0.4 μm and, within the range 0.4 to 2.5 μm, the only significant radiation is solar.

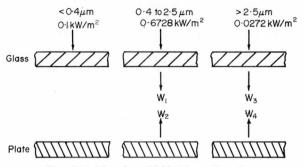

FIG. 5.11. Flat plate shielded by a glass sheet.

0·4 to 2·5 μm:

$$W_1 = 0\cdot9 \times 0\cdot6728 + 0\cdot1 W_2$$

$$W_2 = 0\cdot05 W_1$$

and $W_1 = 0\cdot6085 \text{ kW/m}^2$

$$W_2 = 0\cdot03042 \text{ kW/m}^2$$

$>2\cdot5 \mu$m:

$$W_3 = 0\cdot9 E_{b3} + 0\cdot1 W_4 \qquad \text{(a)}$$

$$W_4 = 0\cdot95 E_{b4} + 0\cdot05 W_3 \qquad \text{(b)}$$

3. *Balances over the whole thermal spectrum.*

Heat absorbed by glass = heat emitted by glass

$$0\cdot9 \times (0\cdot0272 + 0\cdot1 + W_4) = 2 \times 0\cdot9 E_{b3},$$

$$E_{b3} = 0\cdot0636 + W_4/2. \qquad \text{(c)}$$

Note: W_2 is not absorbed at all.

Heat absorbed by plate = heat emitted by plate

$$0\cdot95 \times (W_3 + W_1) = 0\cdot95 E_{b4},$$

$$E_{b4} = 0\cdot6085 + W_3. \qquad \text{(d)}$$

Substituting the values of E_{b3} and E_{b4} in eqs. (a) and (b)

$$W_3 = 0\cdot835 \text{ kW/m}^2$$

and
$$W_4 = 1\cdot413 \text{ kW/m}^2.$$

Equations (c) and (d) give

$E_{b3} = 0\cdot77 \text{ kW/m}^2$ and the glass temperature $= 341$ K (68°C)

and $E_{b4} = 1\cdot444 \text{ kW/m}^2$ and the plate temperature $= 399$ K (126°C).

5.2.3. *Effects of the Earth's Atmosphere*

1. *Cloudless conditions*

Three physical processes affect the attenuation of solar radiation by the atmosphere. They are scattering by the permanent gases in the atmosphere, scattering by aerosols and absorption by water vapour. Attenuation by the atmosphere may be expressed either in the form of eq. (4.1) which was used to describe attenuation by furnace gases or in the form given in eq. (1.21) which is commonly used in solar radiation work. The monochromatic versions of these expressions are

$$G_\lambda = G_{0\lambda}e^{-\alpha_\lambda m} \qquad (5.20)$$

and
$$G_\lambda = G_{0\lambda}t_{a\lambda}{}^m. \qquad (5.21)$$

Over the whole spectrum

$$G = \int_0^\infty G_{0\lambda}e^{-(m\alpha'_1 + m\alpha'_2 + \alpha'_3)}d\lambda \qquad (5.22)$$

$$= \int_0^\infty G_{0\lambda} \times 10^{-(m\alpha_1 + m\alpha_2 + \alpha_3)}d\lambda \qquad (5.23)$$

where α_1 refers to scattering by air molecules, α_2 to scattering by aerosols (dust and droplets) and α_3 to absorption by water vapour, carbon dioxide, ozone, etc.

The variation of these coefficients with wavelength has been discussed in detail by Schuepp (1966). A mean coefficient for pure air over the whole thermal spectrum can be defined by

$$\int_0^\infty G_{0\lambda} \times 10^{-m\alpha_1}d\lambda = G_0 \times 10^{-\alpha_m m}. \qquad (5.24)$$

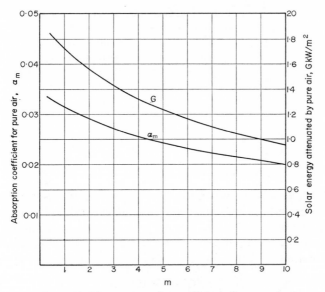

FIG. 5.12. Mean attenuation coefficients for pure air.

This coefficient (α_m) is a function of m and Fig. 5.12 shows the values presented by Robinson (1966) together with values of the attenuated radiation obtained from

$$G = G_0 \times 10^{-\alpha_m m}. \tag{5.25}$$

A mean extinction coefficient for the turbid atmosphere (containing dust, droplets and water vapour) is obtained by multiplying the coefficient for pure air by mT where T is a turbidity factor:

$$G_m = G_0 \times 10^{-\alpha_m m T} \tag{5.26}$$

and

$$T = \log (G_0/G_m)/m\alpha_m \tag{5.27}$$

where G_m is the solar radiation attenuated by passage through the turbid atmosphere.

A large number of turbidity factors have been determined for a range of locations and altitudes and some of these (Foitzik and Hinzpeter, 1958) are given in Table 5.1.

TABLE 5.1. TURBIDITY FACTORS

Location	T (yearly average)
Florence	3·8
Helsinki	2·7
Kew	4·6
Madrid	3·4
Paris	4·0
Rio de Janiero	3·5
Warsaw	3·5

Only direct solar radiation has been discussed so far. Solar radiation also arrives on the Earth as a result of scattering (diffuse sky radiation or D radiation) or reflection from the ground and surrounding surfaces. The scattering functions are complex but a relatively simple expression for estimating the diffuse sky radiation is due to Albrecht (1951):

$$D_z = k_z (G' - G) \cos (z) \qquad (5.28)$$

where k_z is an empirical coefficient, G' is the direct solar radiation in absence of scattering by molecules and aerosols and G is the direct solar radiation when scattering effects are taken into account. For a pure atmosphere, without secondary scattering (and, therefore, independent of reflected radiation from the Earth), k_z should be 0·5.

G' can be obtained by subtracting two factors $(G_{mw} - \Delta_{mO_3})$ related to the amounts of water vapour (mw) and ozone (mO_3) in the atmosphere where m is the relative air mass and w and O_3 are the amounts of water vapour measured as mm of precipitable water and mm of ozone respectively. These data, given by Schuepp (1966), are plotted in Figs. 5.13 and 5.14.

The sum of the vertical components of the direct solar radiation and the diffuse sky radiation is called the global radiation (G_g):

$$G_g = G \cos (z) + D_z. \qquad (5.29)$$

EXAMPLE 5.9

Determine the quantities of direct and diffuse solar radiation under clear sky conditions when the Sun is at its zenith and the total water

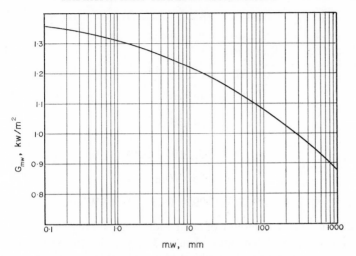

FIG. 5.13. Data for calculating diffuse sky radiation.

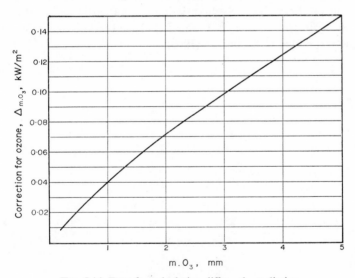

FIG. 5.14. Data for calculating diffuse sky radiation.

content of the atmosphere is 10 mm of precipitable water. The turbidity factor is 2·5 and the amount of ozone is 0·9 cm.

From eq. (5.23) and Fig. 5.12:

$$G = G_0 \times 10^{-2\cdot5 \times 0\cdot0431}$$

$$= 1\cdot39/1\cdot28$$

$$= 1\cdot09 \text{ kW/m}^2$$

$$D = 0\cdot5(1\cdot19 - 1\cdot09)$$

$$= 0\cdot05 \text{ kW/m}^2$$

and $$G_g = 1\cdot09 + 0\cdot05 = 1\cdot14 \text{ kW/m}^2.$$

k_z depends on the amount of solar radiation reflected from the Earth and its atmosphere. The ratio of the radiation reflected to that incident is called the albedo and it is subject to considerable variation as a result of changes in the Earth's surface (vegetation, soil, sea, etc.) and the amount of cloud cover. Such variation cannot easily be described by means of simple formulae but the effects have been discussed in detail in the text edited by Robinson (1966).

CHAPTER 6

Measurement of Radiation and Temperature

THE calculations of radiative heat transfer discussed in previous chapters of this text require data which can be obtained from the literature and/or from experimental measurement. The literature has already been cited; the experimental data can be obtained by using the following techniques.

6.1. MEASUREMENT OF RADIATION

Radiation can be measured over the whole thermal range or over selected wavelength ranges. An instrument which has substantially the same response to all wavelengths is called a total radiation radiometer whilst those instruments which are designed to measure radiation over very small wavelength intervals are known as spectro-radiometers. In either case, a radiometer can be considered to comprise four components:

1. a detecting element which converts the radiant energy to electrical energy;
2. an optical system which ensures that radiation from the target either falls or is focused on the detecting element;
3. an optical system which selects only radiation of the required wavelength range for detection;
4. an amplifier and recorder or indicator which displays the output of the detector.

Such an instrument can then be calibrated with regard to (1) response to known amounts of incident radiation, (2) field of view, (3) speed of response, (4) spectral response and (5) noise level.

6.1.1. *Detecting Elements*

The first requirement of an instrument designed to measure thermal radiation is a detecting or sensing element which produces an output which changes monotonically with the amount of radiation incident upon it. The detector converts radiation incident upon it to a more conveniently measurable form of energy, usually electrical. A number of different types of detector can be used but some are sensitive to radiation over all parts of the thermal spectrum, others to that over a specific range of wavelengths. Basically, there are two types of detectors, thermal and quantum.

6.1.1.1. *Thermal detectors*

Thermal detectors are based on the heating of a detector element by the incident radiation. The detectors are usually thermocouples either used singly or connected in series to form a thermopile. The thermocouple junctions are coated with platinum black in order to raise their absorptivity to a value near 1. A second type of thermal detector is the bolometer which makes use of a material whose conductivity changes significantly with temperature. Platinum was used originally but, more recently, semiconductor materials have constituted the sensing element.

Although thermal detectors are used mainly in the infra-red region of the spectrum their response, because it is brought about by heating alone, is virtually independent of the wavelength of thermal radiation incident upon them. In practice, the use of a window or lens and even the coating on the thermopile or bolometer element means that a real thermal detector does not have a constant response over the whole of the thermal range.

The time constant for a thermocouple detector is of the order of 1 second.

6.1.1.2. *Quantum or photon detectors*

Quantum detectors are based on the interaction of radiation with electrons in a solid which causes them to be excited to a higher energy state and are usually constructed of semiconductor materials. This interaction is utilized in three different types of elements, photovoltaic, photoconductive and photoemissive detectors. In both the photovoltaic and photoconductive elements the excitation manifests itself as changes in the electrical properties of the semiconductor. The phenomenon can be explained in terms of the quantum nature of radiation.

According to the quantum theory, radiation of wavelength λ is composed of particles, known as photons or quanta, each of which possesses energy proportional to $1/\lambda$. For a given semiconductor, a minimum photon energy is required to cause excitation, each photon exciting one electron. The minimum energy corresponds to a critical wavelength (λ_c) above which the radiation cannot be detected. Further, since the energy of each photon is inversely proportional to the wavelength, then for a given amount of energy, the number of photons increases with wavelength. Since the output from a quantum detector is determined by the number of electrons excited, the ideal response of a quantum detector takes the form shown in Fig. 6.1.

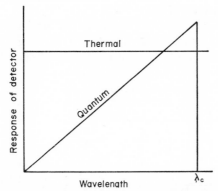

FIG. 6.1. Responses of ideal thermal and quantum detectors.

1. *Photovoltaic cells*

This type of detector is the commonly described photocell. A photovoltaic cell consists of a semiconductor, usually selenium, deposited on a conductor, like iron. Exposure to radiation of a certain wavelength causes a current to flow across the interface between the semiconductor and the conductor.

Photovoltaic cells are essentially sensitive to radiation in the visible region of the spectrum. The time constant of a selenium cell is about 1 millisecond.

2. *Photoconductive cells*

Photoconductive detectors are semiconductors the conductivity of which changes when radiation is incident upon them. An external d.c. voltage is applied across the detector, the current increasing as more radiation is incident upon it. Silicon and germanium detectors are sensitive up to 1 μm and 1·8 μm, respectively, and used in many applications involving a tungsten light source. In addition to excitation of electrons by incident radiation, the thermal energy of the crystal electrons produces a rapidly fluctuating output (noise). This noise can be minimized by operating the detector at low temperatures, for example, indium antimonide detectors may be operated at the temperature of liquid nitrogen (77 K).

The various semiconductor materials used in photoconductive cells are chosen to allow measurement of selected ranges of wavelength. Figure 6.2 shows the response of a number of detector materials to

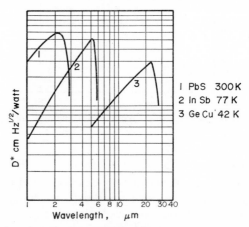

FIG. 6.2. Detectivity of lead sulphide, indium antimonide and germanium/copper detectors.

radiation in the thermal range. The noise equivalent power (NEP) is the minimum amount of radiation which produces a signal to noise ratio of unity. The detectivity of the detector used for measuring radiation is defined as 1/NEP. In order to compare detectors of different areas and response at a particular wavelength, the detectivity is expressed per unit band width and unit area. In some cases, the NEP is proportional to the square root of the area and the normalized detectivity (D^*) quoted for a particular bandwidth of the measuring amplifier is

$$D^* = (\text{area}^{1/2} \times \text{frequency}^{1/2})/\text{NEP}$$

and the units adopted are cm $\text{Hz}^{1/2}$/W.

Time constants of photoconductive cells can vary from about 0·1 microsecond to 0·1 millisecond.

3. *Photoemissive detectors* (*phototubes or photomultipliers*)

In these detectors, the energy provided by radiation causes electrons to be sufficiently excited to leave the solid surface of the detector, a minimum energy being required as in the case of other quantum detectors.

In a photoemissive device, a cathode coated with a photoemissive material emits electrons when radiation of the appropriate wavelength is incident upon it. These electrons are accelerated to a secondary plate by maintaining the plate at a different potential from the cathode. Secondary emission of electrons from this plate, the accelerating plate or dynode, takes place with an increase in the number of electrons. A series of dynodes maintained at a potential gradient by means of a suitable resistor chain produces a much greater final flow of electrons, an effect known as a cascade. The electrons are finally collected at an anode.

The response of a photomultiplier is usually plotted as a function of wavelength in terms of the quantum efficiency η of the cathode (Fig. 6.3). The quantum efficiency is the number of photoelectrons emitted from the cathode per incident photon. On this basis, the cathode sensitivity (mA/watt) is $\lambda \eta e/hc$, where e is the charge on the electron (1·6 × 10^{-19} coulombs), h is Planck's constant (6·62 × 10^{-34} J/s) and c is the velocity of light (3 × 10^8 m/s). The subsequent gain produced by the dynodes varies between 10^3 and 10^6 depending on the applied voltage and the number of dynodes.

The response times of photomultipliers are of the order of 0·1 microsecond.

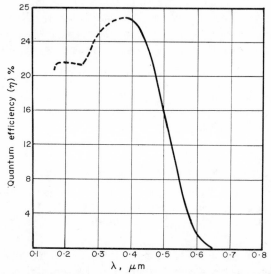

Fig. 6.3. Quantum efficiency of a potassium/caesium photocathode.

6.1.2. *Focusing of Radiation on to the Detecting Element*

Before radiation is incident upon the detecting element, it may be focused by a lens or merely pass through a protecting window. The prime consideration here is the transmissivity of the material comprising the lens or window at the wavelengths of interest and Fig. 6.4 shows the transmissivity of a number of commonly used materials plotted as a function of wavelength. Holter *et al.* (1962) have described the physical properties of a large number of materials used in infra-red work.

6.1.3. *Selection of Wavelength Range by Optical Components*

In order to select the wavelength range of radiation which is required to fall upon the detector, itself sensitive to radiation of certain wavelengths, a spectroscope or filter is used.

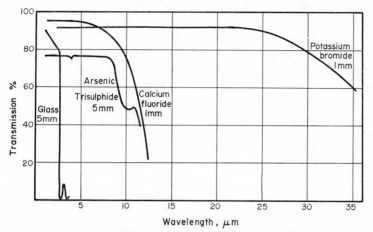

FIG. 6.4. Transmission characteristics of window and lens materials.

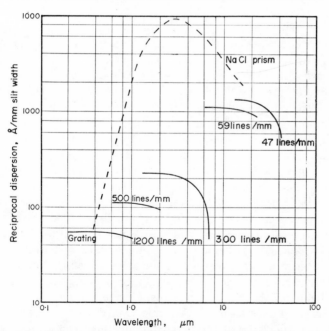

FIG. 6.5. Dispersion produced by a sodium chloride prism and reflection gratings.

6.1.3.1. *Spectroscope*

The spectroscope used in the measurement of thermal radiation is essentially similar to that used in optical work. The core of the instrument is a prism or grating which disperses the radiation spatially. Figure 6.5 shows the dispersion for a prism and a number of gratings indicating that more than one grating has to be used to cover a wavelength range which may be covered by a single prism. One of the disadvantages of the normal diffraction grating is that it disperses the energy incident upon it over a large number of orders. Filters can be used in conjunction with a grating instrument to ensure that the detector receives only diffracted radiation of one order.

The drawback of the prism instrument is that the dispersion changes markedly with wavelength. In addition, the prism material must be transparent over the range of wavelengths being examined.

6.1.3.2. *Filters*

A filter allows radiation of certain wavelengths to be filtered from the radiation incident upon it. Band pass filters transmit radiation over a narrow band of wavelengths and are normally of the interference type. Figure 6.6 shows the characteristics of a narrow band interference filter. Filters of this kind are used instead of a spectroscope when higher energy transmission is desired and lower resolution can be tolerated. The increased energy arises because of the aperture limitation in spectroscopes and the necessity of focusing on to a relatively small slit. Depending on design, the amount of energy transmitted by a filter is between 5 and 500 times that obtained with a spectrometer.

Long-wave pass filters transmit radiation of wavelengths greater than a particular wavelength and are usually constructed of semiconductor materials. These materials absorb radiation up to the critical wavelength, radiation which is utilized in exciting electrons to a higher energy state as described in the section dealing with quantum detectors. Above the critical wavelength, the material transmits radiation. Figure 6.7 shows a typical response curve.

The high absorption up to the critical wavelength makes this type of filter particularly useful in eliminating higher-order spectra of unwanted wavelengths when grating spectroscopes are employed.

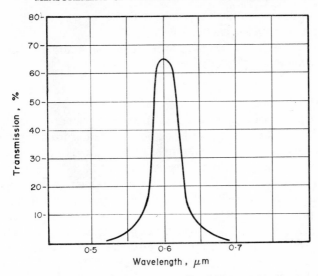

FIG. 6.6. Transmission characteristics of a narrow band interference filter.

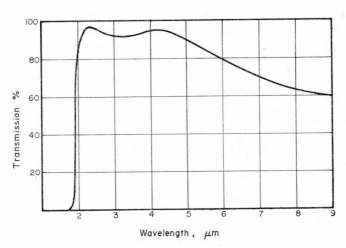

FIG. 6.7. Transmission characteristics of a germanium filter.

A short-wave pass filter transmits radiation of wavelengths smaller than a particular wavelength.

Other devices which are used to make filters are discussed in some detail by Smith *et al.* (1968).

EXAMPLE 6.1

Two narrow band pass filters centred on 3·5 and 4·5 μm with a band width of 1·0 μm receive black-body radiation from a source of area 100 mm² and at a temperature of 1000 K. If the fraction of energy emitted by the source which is received by the filters is 0·001 and the losses through the optical system comprise (a) reflectance losses at three mirrors each of 0·95 reflectivity and (b) transmission losses of 50 and 40 per cent through the 3·5-μm and 4·5-μm filters respectively, calculate the amount of energy incident upon the detector in each case.

The spectral response of the detector varies with wavelength so that the output is 45 per cent of the peak value at 3·5 μm and 60 per cent at 4·5 μm. Calculate the ratio of the outputs from the detector at these wavelengths.

The energy emitted by a black body at a temperature of 1000 K and within the wavelength ranges of 3·0 to 4·0 μm and 4·0 to 5·0 μm can be obtained from Fig. 1.5 or from the data presented by Pivovonsky and Nagel (1961), some of which are given in Appendix 4.

The radiation emitted at wavelengths below 3·0, 4·0 and 5·0 μm, expressed as fractions of the total radiation emitted, are 0·274, 0·481 and 0·634 respectively. Consequently, the energy within the two wavelength ranges centred on 3·5 and 4·5 μm are:

$$3·5 \ \mu\text{m}: 0·208 \times 56·7 = 11·8 \ \text{kW/m}^2,$$

$$4·5 \ \mu\text{m}: 0·153 \times 56·7 = 8·67 \ \text{kW/m}^2.$$

From a source 100 mm² in area:

$$3·5 \ \mu\text{m}: 1·18 \ \text{watts},$$

$$4·5 \ \mu\text{m}: 0·867 \ \text{watts}.$$

A fraction of this energy (0·001) is received by the optical system and the losses in the latter result in the following quantities of energy falling upon the detector:

3·5 μm: $1·18 \times 0·953 \times 0·5 \times 0·001 = 5·06 \times 10^{-4}$ watts,

4·5 μm: $0·867 \times 0·953 \times 0·6 \times 0·001 = 4·46 \times 10^{-4}$ watts.

Ratio of outputs from the detectors (0·35 μm : 0·45 μm)

$$= (5·06 \times 0·45)/(4·47 \times 0·60)$$
$$= 0·851.$$

6.1.4. *Amplification and Display*

All measuring instruments have an indicator or display device which, in the case of electrical instruments, is commonly a moving-coil meter. The galvanometer, a highly sensitive moving-coil meter, can be used but since it is very susceptible to damage from shock or vibration, it is more usual to use an amplifier with a less sensitive moving-coil meter or a chart recorder for measuring signals from detectors.

The output from the detector is a d.c. signal, and if this is of the order of microvolts, as with some solid state detectors or with thermocouples or bolometers exposed to low thermal radiation intensities, difficulties may arise in amplifying the signal. Firstly, d.c. amplifiers are subject to drift and the cost of a detector amplifier with sufficiently low drift can be high. Secondly, the detector itself may generate noise. Both these problems can be much reduced by modulating the signal, that is, converting it to a.c. and using an a.c. amplifier. The a.c. amplifier is not subject to drift and much of the noise can be eliminated if the amplifier is tuned to the modulation frequency. The signal can be modulated either by using a mechanical chopper to modulate or chop the radiation falling on the detector or by electronically modulating the d.c. signal from the detector.

The effect of thermal noise can be further reduced by using a synchronous detector to mix the chopped signal with another noise-free signal modulated at the same frequency. The output from the synchronous detector is a signal comprising the sum and difference frequencies of the original signals. Since the noise is not modulated, this will not appear and, if the higher frequencies are filtered, then the difference frequency, which is a d.c. signal, can be used to drive a meter or a chart recorder.

6.1.5. *Types of Radiometer*

6.1.5.1. *Total radiation radiometers*

1. *Thermocouple radiometers*

The simplest type of radiometer consists essentially of a thermocouple which is placed in the position on which radiation is incident, for example, at the surface of a furnace wall. One version consists of a circular disc of constantan foil, 0·03 mm thick, to the centre of which is soldered a thin (0·03 mm) copper wire. A cylindrical copper block soldered to the outside of the disc acts as a heat sink for the radiation which falls upon the disc. The performance of the radiometer is identical to that of a copper constantan thermocouple measuring the difference in temperature between the centre of the disc and the block. The radiometer is calibrated by means of a standard black-body furnace (Section 6.1.5.3). In order to avoid any convective heat transfer from the surrounding atmosphere to the disc a hemispherical shield, transparent to thermal radiation, may be mounted immediately in front of the constantan disc.

In other kinds of total radiation radiometers, the radiant energy is focused on to the detecting element by means of a lens or mirror system, the lens being constructed of a material, like arsenic trisulphide, which is transparent to thermal radiation.

2. *Ellipsoidal radiometer*

The ellipsoidal radiometer (Fig. 6.8) measures the total thermal radiation incident upon an element of a plane surface.

All radiation falling on the small circular orifice O is focused by the ellipsoidal mirror, which has a gold-plated surface, on to a solid hemisphere (P). The surface of the hemisphere is blackened to absorb practically all the radiation incident upon it, the heat being conducted from the hemisphere to the cooled metal mass M. Constantan wires soldered at points A and B, together with the stainless steel cylinder AB, form a thermocouple which, under steady-state conditions, produces a voltage proportional to the heat received by the hemisphere.

The ellipsoid is purged with dry nitrogen in order to avoid contamination by the combustion gases and particles. Calibration is carried out by means of a black-body furnace (Section 6.1.5.3).

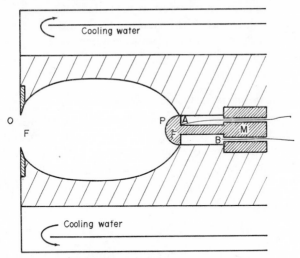

Fig. 6.8. Ellipsoidal radiometer.

The response time of the ellipsoidal radiometer is of the order of 1 minute and does not, therefore, allow rapid changes in incident radiation to be followed. Extensive use of this type of instrument has been made by the International Flame Research Foundation (Chedaille and Braud, 1971).

3. *Heat-flow meter*

This device consists essentially of a relatively long uniform rod, one end of which is cooled to a constant temperature. The other end receives the heat to be measured. It can be used to measure all the heat transferred to the end but, when suitably shielded or if there is minimal convection, only the heat transferred by radiation is measured. A copper rod was used in the version described by Cookson *et al.* (1965). Single constantan wires were soldered into the rod at various positions along its length. A copper tube whose end area was the same as that of the rod surrounded the rod. Since the heat incident on the tube is the same as that incident on the rod they will rise to roughly the same temperature. By maintaining the tube at approximately the same temperature as the rod in this way, radial heat conduction through the rod is minimized and conduction of heat along the rod is essentially one-dimensional. In these circumstances, the heat conducted along the rod (estimated from the thermocouple readings) equals that incident upon the end of the rod.

6.1.5.2. *Spectro-radiometers (spectrometers)*

Apart from those used to determine temperature from the measurement of radiation at a particular range of wavelengths, the spectro-radiometer usually consists of a spectroscope or filter system coupled with a detector, these items being bought independently.

6.1.5.3. *Calibration*

Radiometers may be calibrated

1. by using a black-body source, the temperature of which is measured by a thermocouple, or
2. by comparing its output with a standard radiometer when they are both measuring the temperature of the same object or surface.

Experimentally, the black-body source is a hollow enclosure of a diffuse material maintained at a uniform temperature. Radiation emanating from a small hole in the enclosure is characterized by an emissivity of very nearly unity. In the simplest analysis, based on the leaving flux or radiosity concept, in which the internal surface is treated as a single surface, an estimate of the effective emissivity of the radiation escaping from the aperture can be easily calculated. The analysis provides a close estimate of the effective emissivity when the ratio of the internal surface area to the aperture area is high.

EXAMPLE 6.2

Calculate the effective emissivity of a black-body source which comprises a spherical enclosure, 100 mm in diameter, with a 10-mm diameter aperture. The enclosure is uniformly heated and its surface may be assumed grey with an emissivity of 0·7.

The effective emissivity of the source is given by dividing the flux emerging from the aperture by the black-body emissive power of the enclosure. The flux emerging from the aperture is equal to the flux incident upon it from within the enclosure. The following calculation, based on the assumption that no radiation enters the enclosure from the surroundings, provides an estimate of the flux leaving the enclosure.

The radiosity equation for the surface (1) of the enclosure is

$$W_1 = \epsilon_1 E_{b1} + \rho_1 W_1 F_{11}$$

$$= \epsilon_1 E_{b1}/(1 - \rho_1 F_{11}).$$

The radiation incident on the aperture (2) is

$$Q_2 = A_1 W_1 F_{12}$$

and the flux is

$$q_2 = A_1 W_1 F_{12}/A_2$$

$$= W_1 F_{21}.$$

Since $F_{21} = 1$, then substituting for W_1,

$$q_2 = \epsilon_1 E_{b1}/(1 - \rho_1 F_{11})$$

and the effective emissivity is

$$\epsilon' = \epsilon_1/(1 - \rho_1 F_{11}).$$

The view factor F_{11} is

$$F_{11} = 1 - F_{12}$$

$$= 1 - A_2/A_1$$

$$= 1 - \pi 10^2/(4\pi 10^4 - \pi 10^2)$$

$$= 0.9975$$

and

$$\epsilon' = 0.7/(1 - 0.3 \times 0.9975)$$

$$= 0.9989.$$

Spherical, cylindrical and conical cavities have been used and detailed designs have been discussed by Gouffe (1963). Values of the effective emissivity for spherical and cylindrical shapes have been provided by

Barber (1971), who also listed those materials normally used in the construction of black-body enclosures coupled with their maximum operating temperatures (Table 6.1).

TABLE 6.1. MATERIALS USED IN THE CONSTRUCTION OF BLACK-BODY
ENCLOSURES

Material	Emissivity		Maximum operating temperature, K
	0·65 μm	Total	
Graphite	0·85	0·5	800 (in air)
			3300 (in argon)
Oxidized stainless steel	0·80	0·79	1600
Oxidized nickel	0·9	0·80	1600
Oxidized nimonic alloy (80Ni 20Cr)	0·9	0·89	1600
Silicon carbide	0·7	0·7	1900

6.2. TEMPERATURE MEASUREMENT

The measurement of the temperature of a medium, whether it be gas, liquid or solid, is effected by the use of a detecting element which is either placed in physical contact with the medium or some distance from it. Thermocouples, resistance thermometers, etc., are examples of the former and methods based on radiation examples of the latter. The intensity of radiation emitted from a grey body is a function of its emissivity and its temperature so that a measurement of radiation alone will not be sufficient to provide a value of the temperature unless the emissivity is known. Of course, if only the brightness or effective black-body temperature of the body is required, a measurement of radiation is sufficient.

6.2.1. Thermocouples

The thermocouple is probably the most common industrial method of measuring temperatures higher than ambient. Errors of measurement arise from a number of factors including the conduction of heat along the thermocouple wires and radiation from the thermocouple bead.

The last is significant when measurements of gas temperature are attempted. The temperature of a bare thermocouple in a gas is different from the gas temperature but if the radiation losses can be estimated, the true gas temperature can be calculated (Example 2.2). The error which arises due to radiation to or from the thermocouple can be minimized either by increasing the convective heat transfer to the thermocouple by sucking the hot gases over it at high velocity, as in the suction pyrometer, or by reducing the amount of heat radiated from it by shielding it.

EXAMPLE 6.3

A shielded, butt-welded thermocouple is used to measure the temperature of a gas flowing in a large duct whose walls are at a temperature of 500 K. The shield is very thin and has a diameter four times that of the thermocouple (Fig. 6.9). The emissivities of the thermocouple and shield are 0·8 and 0·3, respectively, and the convective heat

Duct wall (2)

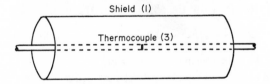
Shield (I)

Thermocouple (3)

FIG. 6.9. Shielded thermocouple.

transfer coefficient to the shield is 0·10 kW/m² K and to the thermocouple is 0·12 kW/m² K. Given that the thermocouple temperature is 800 K, calculate the gas temperature. Compare the thermocouple temperature with that in the absence of the shield.

Since the duct is much larger than the shield, then, according to Section 2.1.3, the net radiative heat transfer from the shield to the duct is

$$Q = A_1 \epsilon_1 (E_{b1} - E_{b2}).$$

The net radiative heat transfer from the shield to the thermocouple can be calculated by the radiosity method:

$$W_1 = \epsilon_1 E_{b1} + \rho_1(W_1 F_{11} + W_3 F_{13}),$$

$$W_3 = \epsilon_3 E_{b3} + \rho_3 W_1 F_{31}.$$

If it is assumed that the lengths of the shield and thermocouple are much greater than their diameters, then

$$F_{31} = 1,$$

$$F_{13} = F_{31} A_3/A_1 = 0 \cdot 25,$$

and
$$F_{11} = 1 - F_{13} = 0 \cdot 75.$$

The radiosity equations can now be written as

$$0 \cdot 475 W_1 - 0 \cdot 175 W_3 = 0 \cdot 3 E_{b1},$$

$$-0 \cdot 2 W_1 \quad + \quad W_3 = 0 \cdot 8 E_{b3}$$

and these may be solved to give

$$W_1 = 0 \cdot 682 E_{b1} + 0 \cdot 318 E_{b3}.$$

The net heat transfer from the shield to the thermocouple is

$$Q = A_1 \epsilon_1 (E_{b1} - 0 \cdot 682 E_{b1} - 0 \cdot 318 E_{b3})/\rho_1.$$

Since the shield is in thermal equilibrium, the total net radiative heat transfer from the shield equals the net convective heat transfer to the shield:

$$A_1 \epsilon_1 (E_{b1} - 0 \cdot 682 E_{b1} - 0 \cdot 318 E_{b3})/\rho_1 + A_1 \epsilon_1 (E_{b1} - E_{b2})$$

$$= 2A_1 \times 0 \cdot 10 \times (T_g - T_1)$$

and
$$T_g = T_1 + 2 \cdot 18 E_{b1} - 1 \cdot 50 E_{b2} - 0 \cdot 681 E_{b3}$$

$$= T_1 + 123 \cdot 6 \times (T_1/1000)^4 - 21 \cdot 1.$$

Since the net heat transfer from the inside of the shield to the thermo-couple is minus that from the thermocouple to the shield, then a heat balance on the thermocouple gives

$$-A_1\epsilon_1(E_{b1} - 0.682E_{b1} - 0.318E_{b3})/\rho_1$$

$$= A_3 \times 0.12 \times (T_g - T_3)$$

and
$$T_g = 4.54E_{b3} - 4.54E_{b1} + T_3$$

$$= 905 - 257(T_1/1000)^4.$$

The two heat balance equations can be solved graphically for T_g to give 808 K.

In the absence of the shield, the heat balance equation for the thermocouple is

$$\epsilon_3(E_{b3} - E_{b2}) = 0.12 \times (T_g - T_3)$$

and
$$T_3 = 831.6 - 378(T_3/1000)^4.$$

A graphical solution gives $T_3 = 726$ K.

Thus, in this case, shielding the thermocouple causes the difference in temperature between the thermocouple and the gas to be reduced from 82 to 8 K.

A further assumption could simplify this calculation a little. If it is assumed that the thermocouple is small compared to the shield, the net heat transfer between the shield and thermocouple can be expressed as $A_3\epsilon_3(E_{b3} - E_{b1})$ and the gas temperature calculated to be 807 K.

6.2.2. *Radiation Pyrometers*

A radiation pyrometer is used when

1. the temperature is too high or the conditions too severe for a sheathed thermocouple to withstand, for example, the roof of an open-hearth steel-melting furnace; or
2. it is difficult or inconvenient to put a thermocouple at the point at which the temperature is required, for example, the temperature of stock moving continuously through a furnace as in the annealing of glass plate and the temperatures within furnaces which are difficult of access, like a cement kiln.

6.2.2.1. *Disappearing filament pyrometer*

The brightness of light from the filament of a calibrated lamp is varied to match the brightness of the light from the hot body whose temperature is to be measured. Figure 6.10 shows the optical arrangement. The image of the hot body or surface (B) is formed in the plane of the filament of the calibrated lamp by the lens O. The filament and the image are viewed through the eyepiece (E) and a filter (F) which usually transmits radiation of wavelength 0·65 μm. When the filament merges, or disappears, into the image of the body, the brightness of the body is judged to be the same as that of the lamp.

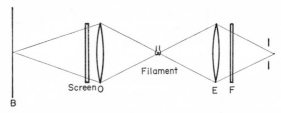

FIG. 6.10. Schematic diagram of a disappearing filament pyrometer.

The filament of the lamp is usually restricted to a maximum temperature of 1400 K. Higher temperatures can be measured by placing an absorbing screen of known transmissivity in front of the objective lens O. In this way, the instrument is usually calibrated for two or even three ranges of temperature to provide an overall range from about 1000 K to 3000 K.

The instrument measures the effective black-body temperature since it is calibrated with a black-body source. The true temperature of the body can be determined from eq. (6.1) if the emissivity of the body is known.

According to the Wien approximation of the Planck equation, the energy emitted by the surface at a wavelength of 0·65 μm is

$$E_{0·65} = \epsilon_{0·65} c_1 (0·65 \times 10^{-6})^5 \exp(-c_2/0·65 \times 10^{-6}T).$$

This corresponds to radiation from a black surface at a temperature T_b:

$$E_{0·65} = c_1 (0·65 \times 10^{-6})^5 \exp(-c_2/0·65 \times 10^{-6}T_b)$$

and

$$T = 1/(4·5 \times 10^{-5} \ln \epsilon_{0·65} + 1/T_b). \tag{6.1}$$

EXAMPLE 6.4

The temperature of a surface of emissivity 0·267 at the red wavelength (0·65 μm) was measured as 1000 K by a disappearing filament pyrometer. Calculate the true temperature of the surface.

$$T = 1/(4·5 \times 10^{-5} \ln(0·267) + 1/1000)$$
$$= 1063 \text{ K}.$$

6.2.2.2. Two-colour method

In this method, the intensity of radiation from a surface is measured at two distinct wavelengths. If the variation of emissivity with wavelength is known or, in its simplest application, the emissivity can be assumed constant with wavelength, the true temperature can be estimated from the following analysis.

Assuming that the Wien approximation to Planck's law may be used without significant loss of precision:

$$E_{b\lambda_1} = c_1 \epsilon_{\lambda_1} \lambda_1^{-5} e^{-c_2/\lambda_1 T}.$$

Comparing the radiation at two wavelengths, λ_1 and λ_2 gives

$$E_{b\lambda_1}/E_{b\lambda_2} = (\lambda_2/\lambda_1)^5 \exp(-c_2[1/\lambda_2 - 1/\lambda_1]/T). \tag{6.2}$$

Substitution of the Wien expression gives

$$T = \frac{(\lambda_2 - \lambda_1)T_1 T_2}{\lambda_2 T_2 - \lambda_1 T_1}. \tag{6.3}$$

EXAMPLE 6.5

Using two narrow band-pass filters, the brightness temperatures of a bed of hot particles were measured at wavelengths of 3·5 and 4·5 μm to be 1180 and 1150 K, respectively. Calculate the true temperature of the bed if it may be assumed that the emissivity is the same at all wavelengths.

From eq. (6.3)

$$T = \frac{(4·5 - 3·5) \times 1180 \times 1150}{4·5 \times 1150 - 3·5 \times 1180}$$
$$= 1360/1·04$$
$$= 1310 \text{ K}.$$

E.C.—F

6.2.2.3. *Total radiation pyrometer*

This type of pyrometer is simply a total radiation radiometer calibrated to read temperature.

6.2.2.4. *Hemispherical pyrometer*

The temperature of a surface can be determined by a method which does not require a knowledge of its emissivity since the temperature is measured under essentially black-body conditions. The radiation detector is mounted on top of the gold-plated reflecting cup which is placed on the hot surface. Figure 6.11 is a diagrammatic sketch of this apparatus for which the following analysis can be applied:

$$\text{Hot surface: } W_1 = \epsilon_1 E_{b1} + \rho_1 W_2 F_{21}, \tag{6.4}$$

$$\text{Hemisphere: } W_2 = \epsilon_2 E_{b2} + \rho_2 W_1 F_{12} + \rho_2 W_2 F_{22}. \tag{6.5}$$

Since $F_{12} = 1$, if surface 1 is flat, $\rho_2 = 1$ and $\epsilon_2 = 0$, then

$$W_2 = W_1 + W_2 F_{22}$$
$$= W_1/(1 - F_{22}) = W_1/F_{21}.$$

Substitution into eq. (6.4) gives

$$(1 - \rho_1)W_1 = \epsilon_1 E_{b1}$$

and

$$W_1 = E_{b1}.$$

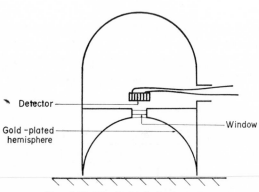

FIG. 6.11. Hemispherical pyrometer.

The radiation leaving surface 1 is that which would be emitted by a black body at that temperature.

6.2.2.5. *Calibration of pyrometers*

Pyrometers may be calibrated using the same techniques as those used for radiometers (Harrison, 1960). In addition, calibration is frequently carried out by measuring the temperature of a standard tungsten strip lamp (Barber, 1971).

6.2.3. *Maximum Intensity of Radiation*

If the surface is grey, then its temperature can be estimated from the shape of the emitted intensity of radiation versus wavelength plot by comparing it with theoretical black-body values. Alternatively, the wavelength at which maximum emission occurs can be used to determine the temperature from Wien's displacement law:

$$\lambda_{max}T = 0.0029 \text{ m K}. \tag{1.7}$$

Equation (1.7) shows that the error in the estimation of temperature is the same as that in locating the wavelength corresponding to maximum emission. Thus, an error of 0.05 μm corresponds to errors in the estimation of temperature of $0.05/\lambda_{max}$.

6.2.4. *Measurements of Flame and Hot Gas Temperatures*

If a radiation pyrometer is sighted on a flame it will receive radiation not only from the flame but also from the background unless, of course, the flame is completely opaque. One method of overcoming this difficulty and which, incidentally, provides an estimate of the emissivity of the flame, was first suggested by Schmidt. The method is more precise when used monochromatically, because the flame is not assumed to be grey, but measurements over the total thermal range may also be appropriate in certain cases.

6.2.4.1. *Schmidt method*

1. *Measurement over the whole thermal range*

The flame and the background source must be assumed grey if the Schmidt method is to be used for the whole thermal range.

The Schmidt method involves three measurements:

(a) radiation from the flame with a cold background (q_1),
(b) radiation from the flame with a hot background (q_2),
(c) radiation from the hot background (q_3).

$$q_1 = \epsilon_f E_{bf}, \tag{6.6}$$

$$q_2 = \epsilon_f E_{bf} + (1 - \epsilon_f) q_3 \tag{6.7}$$

where $E_{bf} = \sigma T_f^4$ and T_f is the temperature of the flame.

Hence $\quad\quad\quad\quad\quad \epsilon_f = 1 - (q_2 - q_1)/q_3 \tag{6.8}$

and $\quad\quad\quad\quad\quad T_f = (q_1/\sigma\epsilon_f)^{1/4}. \tag{6.9}$

2. *Monochromatic Schmidt method*

The basis of this method is essentially the same as that for the whole thermal range. In this case, however, the assumption that the flame is grey need not be invoked. Tourin (1966) has discussed its application in detail and Lowes and Newall (1971) have shown the appropriateness of using the monochromatic variant of this method when small particles are present in a flame, that is, when the absorption coefficient of the flame is inversely proportional to the wavelength (Section 4.1.2(2)).

6.2.4.2. *Two-path method*

A variant of the Schmidt method involves the use of a mirror as a background. Two measurements of radiation are made:

(1) radiation from the flame itself and
(2) radiation from the flame plus radiation from the flame which has been reflected by the mirror and attenuated by the flame.

EXAMPLE 6.6

The temperature of a luminous flame is measured with an optical pyrometer. A mirror, with a reflectance of 0·9 at the appropriate wavelength, is placed behind the flame and a second reading taken. If the measured temperatures are 1800 and 1850 K respectively, calculate the true flame temperature. Assume the flame to be grey.

Radiation from the flame without the mirror:

$$q_1 = \epsilon_f E_{bf}.$$

Radiation from the flame with the mirror:

$$q_2 = \epsilon_f E_{bf} + 0\cdot9(1 - \epsilon_f)\epsilon_f E_{bf}.$$

From these equations, the emissivity is

$$\epsilon_f = (1\cdot9q_1 - q_2)/0\cdot9q_1$$

and

$$E_{bf} = q_1/\epsilon_f = 0\cdot9q_1/(1\cdot9 - q_2/q_1).$$

But

$$q_1 = c_1\lambda^{-5}e^{-c_2/\lambda T}$$

$$= \frac{3\cdot74 \times 10^{-19} \times 10^{30}}{0\cdot65^5 \times 2\cdot179 \times 10^5}$$

$$= 1\cdot467 \times 10^7 \text{ kW/m}^3,$$

$$q_2 = 2\cdot05 \text{ kW/m}^3,$$

$$E_{bf} = 0\cdot9 \times 1\cdot467 \times 10^7/(1\cdot9 - 1\cdot394)$$

$$= 2\cdot62 \times 10^7 \text{ kW/m}^3.$$

From $T = c_2/\lambda \ln(c_1/\lambda^5 E_{bf})$, the temperature is 1889 K.

6.2.4.3. Line-reversal method

When a sodium salt, for example sodium chloride, is introduced into a flame, radiation emitted from the flame at the two yellow D-lines at $0\cdot5890$ μm and $0\cdot5896$ μm. Kirchhoff's law shows that, when light from a background source giving a continuous spectrum is passed through a flame containing sodium vapour, the sodium lines will appear either

1. in absorption as dark lines against the continuous spectrum if the brightness temperature of the background is higher than the flame temperature, or
2. as bright lines—brighter than the continuous spectrum if the brightness temperature is lower.

When the brightness temperature of the background and the flame temperature are the same, then the lines are invisible, that is, they blend perfectly well with the continuous spectrum. Consequently, the flame temperature may be determined by varying the brightness of the background source until the sodium lines disappear (reversal point) and then measuring the brightness temperature of this background with an optical pyrometer.

Thus, the method consists of a comparison between the radiation at the sodium wavelengths and at wavelengths near the sodium wavelengths.

1. *Flame free from particles*

Viewing the background through the flame:

(i) Radiation near the sodium wavelengths,

$$q_1 = c_1 \lambda^{-5} e^{-c_2/\lambda T_b} = W(T_b) \qquad (6.10)$$

where T_b is the brightness temperature of the background and $W(T_b)$ is the Wien approximation to Planck's expression.

Note that, at these wavelengths, the flame is transparent and only radiation from the background source is incident upon the detector.

(ii) Radiation at the sodium wavelengths,

$$q_2 = W(T_b)e^{-K_s L} + W(T_f)(1 - e^{-K_s L}) \qquad (6.11)$$

where K_s is the absorption coefficient at the sodium wavelengths due to the presence of sodium, L is the thickness of the flame and T_f is the flame temperature.

At the reversal point:

$$q_1 = q_2,$$
$$W(T_b) = W(T_f)$$

and

$$T_b = T_f.$$

2. *Particles present in the flame*

The basic experimental procedure is the same for flames containing particles as those without particles. However, the analysis shows the necessity for a further measurement.

(a) Viewing the background through the flame:

(i) Radiation near sodium wavelength,

$$q_1 = W(T_b)e^{-K_pL} \tag{6.12}$$

where K_p is the absorption coefficient due to the presence of particles.

(ii) Radiation at the sodium wavelengths,

$$q_2 = W(T_b)e^{-(K_s + K_p)L} + W(T_f)(1 - e^{-(K_s + K_p)L}). \tag{6.13}$$

At reversal, $q_1 = q_2$ and, if $K_s \gg K_b$,

$$W(T_f) = W(T_b)e^{-K_pL}. \tag{6.14}$$

For $K_pL < 0.2$

$$e^{-K_pL} = 1 - K_pL \tag{6.15}$$

and

$$W(T_f) = W(T_b)(1 - K_pL). \tag{6.16}$$

The fraction of radiation transmitted near the sodium wavelengths (K_p) is obtained by the further measurement of the radiation from the background alone.

(b) Viewing background alone:

$$q_3 = W(T_b) \tag{6.17}$$

and

$$q_1/q_3 = e^{-K_pL}. \tag{6.18}$$

The experimental techniques involved in measuring flame temperatures have been discussed in detail by Gaydon and Wolfhardt (1967) and Tourin (1966).

APPENDIX 1

Kirchhoff's Law

CONSIDER a body at temperature T_1, within a black-body enclosure at temperature T_2.

The radiation absorbed by the body is

$$q_a = \int a_\lambda E_{b\lambda 2} d\lambda \tag{A1.1}$$

and the radiation emitted by the body is

$$q_e = \int \epsilon_\lambda E_{b\lambda 1} d\lambda. \tag{A1.2}$$

If the body is in thermal equilibrium with the enclosure,

$$q_a = q_e$$

or

$$\int a_\lambda E_{b\lambda 2} d\lambda = \int \epsilon_\lambda E_{b\lambda 1} d\lambda. \tag{A1.3}$$

Since

$$T_1 = T_2$$

$$E_{b\lambda 1} = E_{b\lambda 2} \tag{A1.4}$$

and therefore

$$a_\lambda = \epsilon_\lambda. \tag{A1.5}$$

This is one statement of Kirchhoff's law. For most materials, it is found that a_λ (and ϵ_λ) is independent of the amount of the incident radiation and, therefore, of the temperature T_2. Consequently the law may be considered valid whether or not thermal equilibrium prevails.

The overall absorptivity and emissivity are defined as

$$a = \int a_\lambda E_{b\lambda 2} d\lambda / \sigma T_2{}^4 \tag{A1.6}$$

$$\epsilon = \int \epsilon_\lambda E_{b\lambda 1} d\lambda / \sigma T_1{}^4 \tag{A1.7}$$

and, in thermal equilibrium,

$$a = \epsilon. \tag{A1.8}$$

For non-equilibrium conditions, this is not generally true unless a_λ is independent of wavelength, that is, unless the body is grey.

For a grey body:

$$a = a_\lambda = \epsilon_\lambda = \epsilon. \tag{A1.9}$$

Monte Carlo Method

THE Monte Carlo method allows a physical process to be represented by the selection of a series of random numbers. The use of random numbers enables standard procedures for determining errors to be adopted, a necessary part of any numerical method.

A function, descriptive of the physical process, is chosen and expressed in a form that is uniformly distributed over the range of interest in the same way as random numbers are uniformly distributed over the range of possible values.

Consider random numbers between 0 and 1. The probability that a number R is less than or equal to a particular value θ is θ, that is,

$$\text{prob}[R \leqslant \theta] = \theta. \qquad (A2.1)$$

This characteristic of a uniformly distributed variable can now be applied to the selection of a point on a surface.

A2.1. SELECTION OF A POINT OF EMISSION ON A PLANE SURFACE

Consider the circular area of unit radius in Fig. A2.1. If a point P at a distance x from the centre is selected randomly then the probability that it will lie in an area dA is $dA/\pi 1^2$ or $2\pi r dr/\pi 1^2$, if the area is expressed in terms of the distance from the centre, that is, the annular area. Thus, the probability that the point will lie within the elemental area defined by dr is

$$\text{prob}[r \leqslant x \leqslant r + dr] = 2\pi r dr/\pi 1^2 = 2r dr \qquad (A2.2)$$

and the probability that x will have a value less than or equal to r is

$$\text{prob}[x \leqslant r] = \int_0^r 2r dr = r^2. \qquad (A2.3)$$

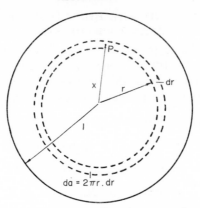

FIG. A2.1. Selection of a point on a surface.

Comparison of eq. (A2.3) with eq. (A2.1) shows that x is not uniformly distributed and cannot be chosen directly from the series of random numbers. However, x^2 is uniformly distributed as can be seen from the following:

From eq. (A2.3):

$$r^2 = \text{prob}[x \leqslant r]$$

$$= \text{prob}[x^2 \leqslant r^2]$$

$$= \text{prob}[R_1 \leqslant r^2] \qquad (A2.4)$$

where R_1 is a random number.

This example illustrates the general rule that the relationship between the process and the selection of random numbers is obtained by an equation of the kind given by eq. (A2.3), that is, the cumulative probability. In general, the probability that a uniformly distributed variable (z) will have a value less than or equal to θ is

$$\text{prob}[z \leqslant \theta] = \frac{\displaystyle\int_{l_1}^{\theta} f(x)dx}{\displaystyle\int_{l_1}^{l_2} f(x)dx} \qquad (A2.5)$$

where l_1 and l_2 are the appropriate limits. z is equivalent to the random number R.

A2.2. SELECTION OF THE DIRECTION OF AN EMITTED BEAM OF RADIATION

The direction of a beam emitted by a surface can be defined in terms of the circumferential (θ) and cone (ϕ) angles and the selection of the appropriate random numbers determined by eq. (A2.5).

From eq. (1.12), the energy emitted from the surface is

$$q = I \cos \phi dw$$

$$= I \cos \phi dA/r^2 \text{ (from Fig. A2.2)}$$

$$= I \cos \phi r d\phi r \sin \phi d\theta/r^2$$

$$= I \cos \phi \sin \phi d\phi d\theta. \tag{A2.6}$$

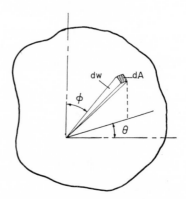

FIG. A2.2. Emission of radiation within a solid angle (dw).

From eqs. (A2.5) and (A2.6):

1. Circumferential angle (θ):

$$\text{prob}[z \leqslant \theta] = \frac{\displaystyle\int_0^\theta I \cos \phi \sin \phi \, d\phi d\theta}{\displaystyle\int_0^{2\pi} I \cos \phi \sin \phi \, d\phi d\theta} \tag{A2.7}$$

$$= \int_0^\theta d\theta/2\pi. \tag{A2.8}$$

Therefore $$z = \theta/2\pi = R_2. \tag{A2.9}$$

2. Cone angle (ϕ):

$$\text{prob}[z \leqslant \phi] = \frac{\displaystyle\int_0^\theta I \cos\phi \sin\phi \, d\phi d\theta}{\displaystyle\int_0^{\pi/2} I \cos\phi \sin\phi \, d\phi d\theta} \qquad (\text{A2.10})$$

$$= \frac{\displaystyle\int_0^\theta \cos\phi \sin\phi \, d\phi}{\displaystyle\int_0^{\pi/2} \cos\phi \sin\phi \, d\phi} \qquad (\text{A2.11})$$

$$= \int_0^\phi \cos\phi \, d(\cos\phi)/[\cos\phi]_0^{\pi/2} \qquad (\text{A2.12})$$

and
$$z = \cos^2\phi - 1 = \sin^2\phi = R_3. \qquad (\text{A2.13})$$

A2.3. NON-DIFFUSE SURFACE

For a non-diffuse surface, I is a function of ϕ and eq. (A2.10) becomes

$$\text{prob}[R_4 \leqslant \phi] = \frac{\displaystyle\int_0^\phi I_\phi \cos\phi \sin\phi \, d\phi}{\displaystyle\int_0^{\pi/2} I_\phi \cos\phi \sin\phi \, d\phi}. \qquad (\text{A2.14})$$

A2.4. VARIATION OF EMISSIVITY WITH WAVELENGTH

If the emissivity of the surface varies with wavelength, then a wavelength may be chosen from

$$\text{prob}[R_5 \leqslant \lambda] = \frac{\displaystyle\int_0^\lambda \epsilon_\lambda E_{b\lambda} d\lambda}{\displaystyle\int_0^\infty \epsilon_\lambda E_{b\lambda} d\lambda} \qquad (\text{A2.15})$$

$$= \frac{\displaystyle\int_0^\lambda \epsilon_\lambda E_{b\lambda} d\lambda}{\epsilon\sigma T^4}. \qquad (\text{A2.16})$$

A2.5. ABSORPTION BY A MEDIUM

The absorption process is modelled by comparing the number of beams absorbed to the total number emitted. A beam is assumed to be completely absorbed after traversing a characteristic length. This length is compared to the dimensions of the enclosure to determine whether the beam is absorbed by the medium or whether it strikes the wall before absorption occurs. The attenuation is e^{-KL} where L is the distance travelled by the beam. A value for L is obtained from

$$\text{prob}[R_6 \leqslant L] = \frac{\int_0^L e^{-KL} dL}{\int_0^\infty e^{-KL} dL} \tag{A2.17}$$

$$= 1 - e^{-KL} = R_6, \tag{A2.18}$$

$$R_7 = 1 - R_6 = e^{-KL} \tag{A2.19}$$

and

$$L = -\ln(R_7)/K. \tag{A2.20}$$

APPENDIX 3

Evaluation of Mean Beam Lengths

THE exchange area between two surfaces A_1 and A_2 is given by eq. (4.41):

$$\overline{s_1 s_2} = \int^{A_1} \int^{A_2} \frac{\cos\phi_1 \cos\phi_2 \, e^{-Kr} dA_1 dA_2}{\pi r^2}. \tag{4.41}$$

The mean beam length L_{12} is defined by eq. (4.44):

$$\overline{s_1 s_2} = A_1 F_{12} e^{-KL_{12}} \tag{4.44}$$

or

$$e^{-KL_{12}} = \int^{A_1} \int^{A_2} \frac{\cos\phi_1 \cos\phi_2 \, e^{-Kr} dA_1 dA_2}{A_1 F_{12} \pi r^2}. \tag{A3.1}$$

If Kr is small,

$$(1 - KL_{12}) = \int^{A_1} \int^{A_2} \frac{(1 - Kr)\cos\phi_1 \cos\phi_2 \, dA_1 dA_2}{A_1 F_{12} \pi r^2} \tag{A3.2}$$

or

$$L_{12} = \int^{A_1} \int^{A_2} \frac{\cos\phi_1 \cos\phi_2 \, dA_1 dA_2}{A_1 F_{12} \pi r}. \tag{A3.3}$$

This equation has been evaluated by Dunkle (1964) for parallel and perpendicular rectangular plates and some of these data are plotted in Figs. A3.1 and A3.2.

Although these data are adequate for calculations of heat transfer in rectangular enclosures where the six bounding planes are treated as separate rectangular surfaces, it is often convenient to treat the four

145

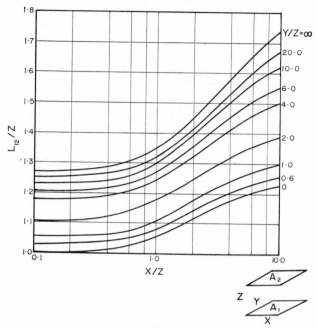

FIG. A3.1. Mean beam lengths for parallel plate geometry.

vertical walls of a furnace as a single surface. In this case, the mean beam lengths can be derived from Dunkle's data.

Consider radiation transmitted from the floor of a rectangular furnace (surface 1) to the four vertical walls (surfaces a, b, c and d) which are to be treated as a single surface (2). The fraction of radiation emitted by surface 1 and incident upon the four vertical walls is

$$F_{12}e^{-KL_{12}} = F_{1a}e^{-KL_{1a}} + F_{1b}e^{-KL_{1b}}$$

$$+ F_{1c}e^{-KL_{1c}} + F_{1d}e^{-KL_{1d}}. \qquad (A3.4)$$

Since

$$F_{12} = F_{1a} + F_{1b} + F_{1c} + F_{1d} \qquad (A3.5)$$

and KL is small,

$$L_{12} = (F_{1a}L_{1a} + F_{1b}L_{1b} + F_{1c}L_{1c} + F_{1d}L_{1d})/F_{12}. \qquad (A3.6)$$

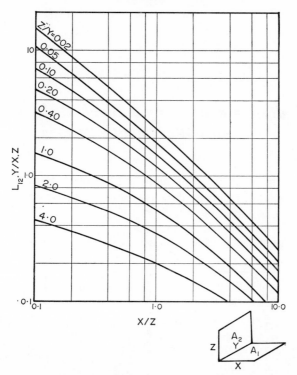

FIG. A3.2. Mean beam lengths for perpendicular plate geometry.

A similar analysis enables view factors like L_{22} to be calculated:

$$L_{22} = \frac{A_a(F_{ab}L_{ab} + F_{ac}L_{ac} + F_{ad}L_{ad}) + A_b(F_{bc}L_{bc} + F_{bd}L_{bd}) + A_cF_{cd}L_{cd}}{A_a(F_{ab} + F_{ac} + F_{ad}) + A_b(F_{bc} + F_{bd}) + A_cF_{cd}}.$$

(A3.7)

The assumption that KL is small can lead to errors in the evaluation of mean beam length as has been discussed by Chamberlain *et al.* (1972).

Radiation emitted by a Black Body within Stated Wavelength Ranges

THE percentages of radiation emitted by a black body in the range 0 to λT, selected from the data presented by Pivovonsky and Nagel (1961), are given in Table A4.

TABLE A4. PERCENTAGES OF RADIATION EMITTED BY A BLACK BODY IN THE RANGE 0 TO λT

λT μm K	D %	λT μm K	D %	λT μm K	D %	λT μm K	D %
1000	0·0323	2800	22·82	6500	77·66	24,000	99·12
1100	0·0916	3000	27·36	7000	80·83	26,000	99·30
1200	0·214	3200	31·85	7500	83·46	28,000	99·43
1300	0·434	3400	36·21	8000	85·64	30,000	99·53
1400	0·782	3600	40·40	8500	87·47	35,000	99·70
1500	1.290	3800	44·38	9000	89·01	40,000	99·79
1600	1·979	4000	48·13	9500	90·32	45,000	99·85
1700	2·862	4200	51·64	10,000	91·43	50,000	99·89
1800	3·946	4400	54·92	12,000	94·51	55,000	99·92
1900	5·225	4600	57·96	14,000	96·29	60,000	99·94
2000	6·690	4800	60·79	16,000	97·38	70,000	99·96
2200	10·11	5000	63·41	18,000	98·08	80,000	99·97
2400	14·05	5500	69·12	20,000	98·56	90,000	99·98
2600	18·34	6000	73·81	22,000	98·89	100,000	99·99

Emissivities of Atmospheres containing Mixtures of Carbon Dioxide and Water Vapour

THE emissivities of atmospheres containing carbon dioxide and water vapour in the ratios 1:1 and 1:2, corresponding to the combustion products of petroleum fuels and methane respectively, are presented in Figs. A5.1 and A5.2 (Hadvig, 1970).

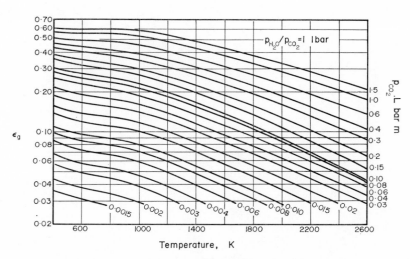

FIG. A5.1. Emissivities of atmospheres containing carbon dioxide and water vapour in the ratio 1:1.

149

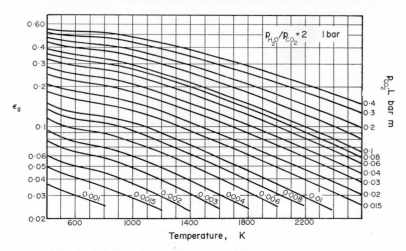

FIG. A5.2. Emissivities of atmospheres containing carbon dioxide and water vapour in the ratio 1:2.

References

ALBRECHT, F. (1951) *Arch. Meteorol. Geophys. Bioklimatol.* Ser. B, **3**, 220.
BARBER, C. R. (1971) *Calibration of Thermometers*, N.P.L./Dept. Trade Industry, H.M.S.O.
BASSETT, C. R. and PRITCHARD, M. D. W. (1970) *Environmental Physics: Heating*, Longmans, London.
BEER, J. M. and JOHNSON, T. R. (1972) 4th *Symp. Flames Industry*, Paper No. 4, Inst. Fuel and Brit. Flame Res. Comm.
BOOTH, F. (1949) *Proc. Phys. Soc.* **62A**, 95.
BRINKWORTH, B. J. (1972) *Solar Energy for Man*, Compton Press, Salisbury.
CHAMBERLAIN, C. T., GRAY, W. A. and MÜLLER, R. (1972) *Silikat J.* **11**, 281.
CHEDAILLE, J. and BRAUD, Y. (1971) *Industrial Flames: 1. Measurements in Flames*, Arnold, London.
COOKSON, R., DUNHAM, P. and KILHAM, J. K. (1965) *J. Scient. Instrum.* **42**, 260.
DEVOS, J. C. (1954) *Physica*, **20**, 669.
DUNKLE, R. V. (1964) *J. Heat Transfer* (ASME), p. 75.
FIELD, M. A., GILL, D. W., MORGAN, B. B. and HAWKESLEY, P. G. W. (1967) *Combustion of Pulverised Coal*, B.C.U.R.A./Inst. Fuel.
FOITZIK, L. and HINZPETER, H. (1958) *Sonnenstrahlung und Lufttrübung*, Geest u. Portig K.G., Leipzig.
GAYDON, A. G. and WOLFHARDT, H.G. (1970) *Flames: Their Structure, Radiation and Temperature*, Chapman & Hall, London.
GOUFFE, A. (1963) *Rev. Gen. Therm.* **19**, 799 and **20**, 925.
HADVIG, S. (1970) *J. Inst. Fuel.* **43**, 129.
HARRISON, T. R. (1960) *Radiation Pyrometry*, Wiley, New York.
HAWKSLEY, P. G. W. (1952) *B.C.U.R.A. Monthly Bull.* **16**, Nos. 4 and 5.
HERDAN, G. (1960) *Small Particle Statistics*, Butterworths, London.
HOLTER, M. R., NUDELMAN, S., SUITS, G. H., WOLFE, E. W. L. and ZISSIS, G. J. (1962) *Fundamentals of Infra-red Technology*, Macmillan, New York.
HOTTEL, H. C. and COHEN, E. S. (1958) *A.I.Ch.E. J.* **4**, 3.
HOTTEL, H. C. and SAROFIM, A. F. (1967) *Radiative Transfer*, McGraw-Hill, New York.
HOWELL, J. R. (1968) *Advances in Heat Transfer*, **5**, 1, Academic Press, New York.
JONES, W. P. (1967) *Air Conditioning Engineering*, Arnold, London.
KREITH, F. (1962) *Radiation Heat Transfer for Spacecraft and Solar Plant*, International Textbook, Scranton.
KREITH, F. (1967) *Principles of Heat Transfer*, International Textbook, Scranton.
LECKNER, B. (1972) *Comb. Flame*, **19**, 33.
LEUENBERGER, H. and PERSON, R. A. (1956) ASME Paper No. 56–A144.
LOVE, T. J. (1968) *Radiative Heat Transfer*, Merrill, Columbus.

151

LOWES, T. M. and NEWALL, A. J. (1971) 37th Autumn Res. Meeting, Inst. Gas Engrs., Comm. No. 862.

MCADAMS, W. H. (1954) *Heat Transmission*, McGraw-Hill, New York.

MCCARTNEY, J. J. and ERGUN, S. (1958) *Fuel*, **37**, 272.

OPPENHEIM, A. K. (1956) ASME Paper No. 54–A75.

PIVOVONSKY, M. and NAGEL, M. R. (1961) *Tables of Black Body Radiation Functions*, Macmillan, New York.

ROBINSON, N. (1966) *Solar Radiation*, Elsevier, Amsterdam.

SAROFIM, A. F. (1962) Sc.D. Thesis, M.I.T.

SCHUEPP, W. (1966) in *Solar Radiation* (ed. ROBINSON, N.).

SMITH, R. A., JONES, F. E. and CHASMER, P. E. (1968) *The Detection and Measurement of Infra-red Radiation*, Clarendon Press, Oxford.

SPARROW, E. M. and CESS, R. D. (1966) *Radiation Heat Transfer*, Brooks/Cole, California.

STEWARD, F. R. and CANNON, P. (1971) *Int. J. Heat Mass Transfer*, **14**, 245.

TANIGUCHI, H. (1969) *Bull. Jap. Scient. Mech. Eng.* **12**, 1.

THRING, M. W. (1962) *The Science of Flames and Furnaces*, Chapman & Hall, London.

TOULUOKIAN, Y. S. (1972) *Thermophysical Properties of Matter*, IFI Plenum, New York.

TOURIN, R. H. (1966) *Spectroscopic Gas Temperature Measurement*, Elsevier, Amsterdam.

TRINKS, W. and MAWHINNEY, M. H. (1961) *Industrial Furnaces*, Wiley, New York.

WIEBELT, J. A. (1966) *Engineering Radiation Heat Transfer*, Holt, Rinehart & Winston, New York.

EFFICIENT USE OF FUEL (1958), H.M.S.O., London.

Glossary

Absorptivity: ratio of the radiation absorbed by a body to that incident upon it.

Albedo: ratio of the radiation reflected from the Earth to the incident radiation.

Altitude: the angle between a line from the Sun to the centre of the Earth and the tangent to the surface of the Earth.

Aphelion distance: the greatest distance between the Sun and the Earth.

Attenuation coefficient: rate of decrease of radiation with distance as it passes through an absorbing medium, expressed per unit quantity of radiation.

Azimuthal angle: the angle between due south and the direction (compass) of the Sun.

Band pass filter: a filter which transmits radiation over a narrow band of wavelength.

Black body or black surface: a body (or surface) which (1) absorbs all radiation incident upon it (and reflects or transmits none) or (2) emits, at any particular temperature, the maximum possible amount of thermal radiation.

Black-body source or enclosure: a source (or enclosure) whose absorption and emission characteristics closely approach that of a black body.

Bolometer: a thermal detector which uses a metal, whose conductivity changes markedly with temperature, as the sensitive element.

Brightness temperature: the temperature of a black body which emits the same amount of radiation as the body being considered.

Celestial sphere: a sphere of infinite radius with the Earth as its centre.

Chopping: modulation of radiation by placing a rotating toothed disc in the path of the beam.

Crossed-string method: a technique for determining exchange areas between long, parallel bodies of uniform cross-section.

Declination: the celestial equivalent of latitude.

Detectivity: reciprocal of the noise equivalent power.

Diffuse reflection: uniform reflection of radiation over all angles, independent of the angle of incidence.

Dispersion: the separation of different wavelengths by a prism or grating.

Dynodes: the secondary emission plates in a photomultiplier or phototube.

Ecliptic axis: the axis, passing through the Sun, around which the Earth revolves.

Ecliptic plane: the plane in which the Earth revolves.

Effective black-body temperature: the same as the brightness temperature.

Emissive power: the radiant energy emitted by a body.

Emissivity: the ratio of the radiant energy emitted by the body under consideration to that emitted by a black body at the same temperature.

153

Exchange area: the ratio of the radiation emitted by surface 1 which is incident on surface 2 to the black body emissive power of surface 1.

Extinction coefficient: the same as the attenuation coefficient.

Flux: the rate of flow of radiation per unit area.

Global radiation: the sum of the vertical component of the direct solar radiation on the Earth and the diffuse (or scattered) solar radiation.

Greenhouse effect: solar heating of objects shielded by glass which transmits solar radiation but absorbs most of the radiation emitted by the bodies themselves.

Greenwich hour angle: the celestial equivalent of longitude.

Grey body: a body whose emissivity is constant with wavelength.

Infra-red region: Electromagnetic radiation characterized by wavelengths of 0.76 to 1000 μm.

Intensity of radiation: radiant energy per unit area per unit time per unit solid angle.

Local hour angle: Greenwich hour angle (GHA) minus longitude west of Greenwich, or GHA minus longitude east of Greenwich.

Long furnace model: a plug flow model of a furnace which is long enough for radiative transfer along its length to be ignored.

Long-wave pass filter: a filter which transmits radiation at wavelengths greater than a critical value.

Mean beam length: an average length describing absorption of radiation by a medium.

Modulation: the modification of a signal by the superimposition of another.

Monochromatic radiation: radiation of one particular wavelength.

Monte Carlo method: a statistical method used in the analysis of thermal radiation whereby the radiation is represented by a large number of beams whose origin and direction are selected from a series of random numbers.

Network method: use of an electrical analogy of heat transfer.

Noise: unwanted signal.

Noise equivalent power: the minimum amount of radiation which produces a signal to noise ratio of unity.

Normalized detectivity: [(area × frequency)$^{1/2}$ divided by detectivity— used to compare responses of different detectors of different areas and at different frequencies.

Perihelion distance: the minimum distance between the Earth and the Sun.

Photocell: photovoltaic cell.

Photoconductive cell: a semiconductor detector which utilizes the changes in electrical conductivity which occur as a result of interaction between radiation and electrons in the material.

Photoemissive cell: a detector which utilizes electron emission from the surface of some materials when radiation is incident upon them.

Photomultiplier: a detector with a photoemissive cell as the sensitive element and dynodes to amplify the signal.

Photon: quantum or bundle of radiation.

Photosphere: the Sun as it appears to an observer on the Earth.

Phototube: a photomultiplier.

Photovoltaic cell: a semiconductor detector with a rectifying junction. Photon–electron interactions produce a voltage across the device.

Plug flow model: a flow characterized by constant transverse temperatures and compositions.

Pyrometer: a radiometer calibrated to read temperature.

Quantum detector: a detector which utilizes the effects of the interaction between radiation and electrons in the detector material.

Quantum efficiency: number of photo-electrons emitted from the cathode per incident photon.

Radiation: (1) transmission of energy by electromagnetic waves; (2) radiant energy.

Radiometer: an instrument which measures radiation.

Radiosity: the sum of the incident and reflected radiation fluxes from a surface.

Reflectivity: the ratio of the radiation reflected by a body to that incident upon it.

Relative air mass: ratio of the actual path length of radiation through the Earth's atmosphere to the shortest possible path length.

Scattering: attenuation of radiation passing through a medium by means other than absorption.

Selective emitter: a material whose monochromatic emissivity is a function of wavelength, angle of emission or surface temperature.

Short-wave pass filter: a filter which transmits radiation at wavelengths smaller than a critical value.

Solar constant: the amount of solar radiation incident upon a surface normal to the radiation and situated just outside the Earth's atmosphere.

Spectral emission: emission at particular wavelengths.

Spectral response: the sensitivity of a detector to radiation of a particular wavelength.

Spectrometer: a radiometer which measures radiation over small intervals of wavelength.

Spectro-radiometer: a spectrometer.

Spectroscope: an instrument which is capable of dispersing radiation.

Specular reflection: reflection of radiation for which the angle of incidence equals the angle of reflection.

Speed of response: an indication of the time taken for a detector to respond; quantitatively given by the time constant.

Thermal detector: a detector which utilizes the heating effect of radiation.

Thermal radiation: radiant energy in the wavelength range $0 \cdot 1$ to $100 \ \mu m$.

Thermopile: a detector comprising a number of thermocouples in series.

Time constant: the time for a detector signal to reach $0 \cdot 63 \ (= 1 - 1/e)$ of its final value.

Total radiation radiometer: a radiometer equally sensitive to radiation over all wavelengths.

Turbid atmosphere: an atmosphere containing suspended particulate material.

Turbidity factor: the factor by which the relative air mass is multiplied to account for attenuation by a turbid atmosphere.

Ultra-violet: radiation characterized by wavelengths in the $0 \cdot 1$ to $0 \cdot 38 \ \mu$ range.

View factor: the fraction of the total radiation emitted by one surface which is directly incident upon a second surface.

Visible region: thermal radiation which affects the optic nerves.

Well-stirred model: a mathematical model in which the reactants and products are assumed to be of uniform temperature and composition.

Zenith angle: the angle between a line from the Sun to the centre of the Earth and the normal to the surface of the Earth (90° —altitude).

Zone: an area of surface or a volume of absorbing medium, which, for the purposes of calculation, is assumed to be isothermal.

Index

157

158

INDEX